# THE IMPACTFUL TECHNOLOGY LEADER

## PURSUE YOUR PURPOSE TO INNOVATE, MOTIVATE AND INSPIRE

## CHRIS CARUSO

THE IMPACTFUL TECHNOLOGY LEADER
Copyright © 2026 by Chris Caruso



For press inquiries, contact the author at:
chris@christopherrcaruso.com

First Edition March 2026

ISBN (print): 979-8-9937455-0-3
ISBN (ebook): 979-8-9937455-1-0

Cover Design: Steve Kuhn
Interior Design: Domini Dragoone
Editor: Kelly Lydick
www.writeeditdesignlab.com

For my wife, my steadfast partner and constant source of love,

and for our sons and their families,

whose lives fill me with pride, hope, and wonder.

You remind me that while the world often

measures impact in achievements,

the truest measure of a life's impact

is the love we give and share.

# Contents


Introduction ................................................................. 1

**Remind Me Again,**
**Why Did I Want to Become a Technology Leader?** ........................... 5

**PART ONE:**
**Impacting the Firm: Driving Innovation through Technology**

CHAPTER 1: **Impacting the Firm: Why It Matters** ..................................... 21

CHAPTER 2: **Impacting the Firm: How Do They Do It?** ........................... 37

    Vishal Gupta (Lexmark) .......................................................37
    *Discovery Skill: Experimenting*

    Fortune 200 CIO ................................................................. 42
    *Discovery Skill: Questioning*

    Chris Caruso (PPG) ............................................................ 48
    *Discovery Skill: Observing*

    Jim Rinaldi (NASA Jet Propulsion Laboratory) ............................55
    *Discovery Skill: Networking*

    Chris Caruso (PPG) ............................................................59
    *Discovery Skill: Associating*

CHAPTER 3: **Integrating All Five Discovery Skills** ..................................65

    PPG's Digital Paint Purchase Process: An Innovation Story ................ 66

CHAPTER 4: **Enabling Innovation by Listening,**
**Communicating, & Acting Courageously (Renee Zaugg)** ..................77

CHAPTER 5: **Impacting the Firm: What Did We Learn** ........................... 87

**PART TWO:**

**Impacting the Team: Creating a Great Workplace**

CHAPTER 6: Impacting the Team: Why It Matters ...................................95

CHAPTER 7: Impacting the Team: How Do They Do It? ......................105

    Steve Zerby (Owens Corning) ..........................................................105

    Graeme Thompson (Informatica) ................................................... 110

    Hans Keller (Erickson Senior Living) ........................................... 118

CHAPTER 8: "If you focus on the people, the technology will follow." ....125

CHAPTER 9: Communicating to Connect (Dick Daniels)....................... 135

CHAPTER 10: Impacting the Team: What Did We Learn .....................143

**PART THREE:**

**Impacting the Profession: Expanding
and Diversifying the Tech Workforce**

CHAPTER 11: Impacting the Profession: Why It Matters..................... 151

CHAPTER 12: Impacting the Profession: How Do They Do It? ............163

    Dr. Philile Mkhize.............................................................................164
    *Lifting as She Leads: A Mission to Empower Girls and Women in Tech*

    Cindy McKenzie .................................................................................173
    *Inspiring University Students to Pursue Careers in Tech*

    Simon Reiter .......................................................................................182
    *Mentoring to Develop the Next Generation of Tech Leaders*

    Julie Ragland ...................................................................................... 190
    *Opening Doors: Expanding Access to Tech Careers*

CHAPTER 13: Doing the Next Right Thing: Unlocking Opportunities with Tech Apprenticeships (Barry Lowry) ...........................................199

CHAPTER 14: Impacting the Profession: What Did We Learn .............211

The Interconnected Journey of Technology Leadership ...................219

Being a Technology Leader, It Really Can Be a Wonderful Life.......227

APPENDIX A ......................................................................................231
Impacting the Firm - Innovation Discovery Skills Survey

APPENDIX B.......................................................................................239
Impacting the Team - Herzberg's Two-Factor Theory Survey

APPENDIX C.......................................................................................251
Impacting the Profession - Deeper & More Diverse Talent Pool Survey

APPENDIX D ......................................................................................259
Impacting the Profession - STEM Career Development Organizations

List of Figures ..................................................................................262

Works Cited ....................................................................................263

Acknowledgments ...........................................................................267

About the Author.............................................................................270

# THE IMPACTFUL

# TECHNOLOGY

# LEADER

# Introduction

bake bread. Some people golf, ski, or fish to unwind. I bake bread. It's a hobby that has nothing to do with my professional life, yet it shares much with what I have learned about being a leader.

My fascination with bread baking began in 1982, on my first trip to Europe. I was traveling through Italy with my father, and after arriving in Rome, we sat down for breakfast and were served a basket of rolls called *Rosetta*, named for their flower-petal shape. They were light and golden, and when I sliced one open, I was amazed to find it mostly hollow inside. It was a simple thing, but I couldn't stop wondering: *How do they do it?*

That question stayed with me as we traveled across the country, trying amazing regional foods and sampling breads I'd never eaten before. When we returned home, I searched for those same breads in the U.S. without success, so I decided I'd learn to bake them myself.

I read books, attended classes, and baked *a lot* of bread. At the beginning, my bread wasn't very good. The more I baked, the more I learned, and the more I began to share the bread with my family, friends, and neighbors. In many ways, it became something I was known for and passionate about.

People would ask me, "Have you tried making sourdough? Do you have a good recipe for gluten-free bread?" Others shared the same struggles I once had, getting the dough to rise and achieving that perfect balance of a crisp crust and a light, airy center. I understood because I'd been there. I've always enjoyed sharing the tips and techniques I've picked up along the way that help novice bakers learn faster, avoid frustration, and build their confidence.

After baking thousands of loaves over several decades, I've come to appreciate a simple truth: most bread begins with just four ingredients: flour, water, yeast, and salt. Whether Italian or French, whole wheat or gluten-free, nearly every recipe starts with the same basic elements. What sets each loaf apart isn't just the ingredients but the proportions, the kneading technique, and the time it's given to rise. It's a chemical process that can transform these simple ingredients into an incredible loaf of bread.

Much like pursuing the perfect loaf, my work as a technology leader was fueled by a similar curiosity to understand what truly sets the best apart. Over the years, I've learned that building innovative and engaged technology teams requires a similar formula with a few essential ingredients: talent, resources, purpose, and culture, that are brought together with care, time, and intention. Exceptional technology leaders understand how these elements combine to create something far greater than the sum of their parts.

I learned this after being involved with teams recognized for innovation with the CIO 100 award from Foundry's CIO and Computerworld's "Best Places to Work in IT" award. While preparing to nominate the organization again for these two industry honors, I noticed a pattern as I scanned the list of winners over several years. Many of the same companies appeared again and again. I asked the same question that had driven my bread-baking curiosity: *How do they do it?*

I wanted to understand what set these organizations apart and why some consistently delivered innovative technology solutions that provided a competitive advantage for their firms, while others built exceptional workplaces where employees thrived. What intrigued me most was that a few standout companies repeatedly appeared on both lists, suggesting a powerful connection between innovation and culture.

Years later, as I entered retirement, that question lingered with me. It became one of the driving reasons I decided to write this book. I knew that the leaders of these organizations were doing something different that should be shared with other current and aspiring technology leaders.

At first, I didn't fully grasp how deeply connected innovation and culture truly are. As I began interviewing some of the "Best Places to Work" leaders for this book, it became apparent that there was a connection between driving innovation and having an outstanding workplace. These leaders understood that to have a sustainable culture of innovation, they also needed a highly engaged and resilient team.

Another key insight from these interviews was that these leaders recognized the importance of accessing a deeper and more diverse talent pool to fuel ongoing innovation initiatives and sustain their exceptional workplaces. They understood that building a great workplace and driving innovation go hand in hand, and both require attracting people with a wide range of perspectives, experiences, and skills.

Accessing this talent required these leaders to get personally involved. They needed to inspire students early in their career journeys and provide meaningful support as they pursued their studies and transitioned into the technology profession.

These leaders didn't stumble upon success by chance. They were intentional about how they built teams, nurtured culture, and drove innovation that mattered. Just like crafting a great loaf of bread, it took patience, experimentation, and a deep understanding of the ingredients that matter most.

This book shares their insights, experiences, and lessons not as a rigid formula, but as inspiration and guidance for your leadership journey. Whether you're building your first team or shaping the direction of an entire organization, the stories and strategies in the chapters ahead are meant to help you lead with greater clarity and purpose.

# Remind Me Again, Why Did I Want to Become a Technology Leader?

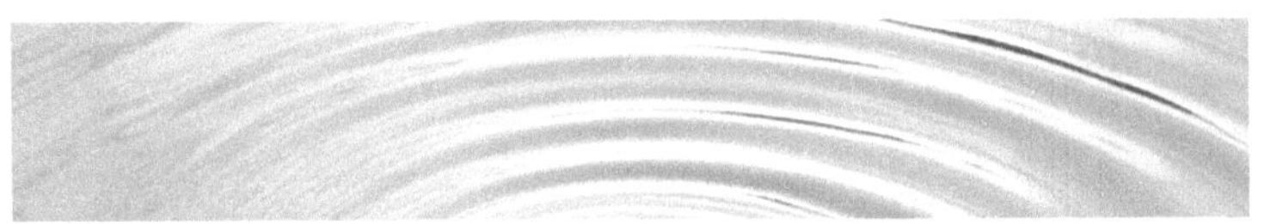

"Be careful what you wish for, lest it come true!"
**—Aesop's Fables**

This timeless warning from Aesop's fables rings especially true in technology leadership. Have you ever taken a moment to consider why you aspire to be a technology leader? Is it simply the next phase in your career journey? Were you motivated to elevate your standing within your organization, among fellow professionals, or with your friends and family? Did you pursue a management position, thinking it would increase your compensation? If any of these apply, you should reconsider your reasons for wanting the job.

For many, being a leader in the technology profession can seem like an exciting and appealing role, filled with lavish dinners, award ceremonies, and the satisfaction of basking in the glow of your team's achievements. Whether you hold this role now or aspire to it in the future, it's essential to understand that positions such as manager, director, or chief information officer are more challenging than they may appear. The following story will help to provide a quick reality check on what it takes to lead others.

## Technology Leadership: Into the Deep End

Dick Daniels, who served as a Chief Information Officer (CIO) in the healthcare and financial services industries, shared the following story that illustrates that being a leader is much more than many expect when they accept the role.

In the early 80s, I began my career in IT as a computer operator and did some programming. During this time, I started aspiring to lead an IT organization. Over the years, I held various management positions, overseeing data centers, networks, technical support, and applications.

At one point, when I was a manager, my office was located next to the person in charge of the entire IT department. Late on a Thursday, around 5:30 p.m., I had just turned off my computer and put some things into my briefcase when someone from the Finance department entered my office. He closed the door, sat down at my office table, and said, "Dick, I need to talk to you." "Tomorrow, we're going to announce that we're promoting you to the head of IT for the company." I was excited. I had been waiting for a promotion like this. Then it occurred to me, "Wait, somebody already has that job in the next office." I asked, "What's happening with the guy in the corner office?" He said, "We're letting him go in the morning." I felt terrible, but I quickly realized that this was the moment I had been waiting for my entire life.

Immediately after telling me that I would be promoted, he informed me that I would need to provide him with twenty-six names of people I would be laying off in the department. He had his pen in his hand, ready to begin writing names. I was sitting there, and all of these emotions began to emerge. Maybe I should give him my badge because I've worked with these people for many years. He could see the difficulty I was going through and said, "Okay, I see you need some time. I'll just come by tomorrow at 8:00 a.m." And he got up and walked out. While I always wanted this position, I didn't realize the job would be like this.

I went back, turned my computer on, printed out the organization charts, and put them in my briefcase. It was a long drive home, and I continued to think, "This is the opportunity I have always wanted, and now that I have the job, this is what it is." So, I got home, pulled out the org charts, and produced twenty-six names. And, sure enough, at 8:00 a.m. the following day, he was there and wanted the names.

The plan was for a significant layoff at the company, and I was responsible for delivering IT's contribution. The announcement of my new role was made on Friday, but I waited until Monday to inform my new direct reports, who were previously my peers, about the layoff. I told them they had nothing to do with selecting the people to lay off and, since they didn't, I wouldn't ask them to deliver the bad news. That week, I laid off all twenty-six of those employees. That was my first week of being in charge of the IT department.

Dick shares this story whenever someone expresses interest in a management position. He advises them to understand the job responsibilities before accepting a leadership role. Despite always wanting to be the head of the department, Dick didn't fully grasp what it would take to fulfill the role. He now cautions anyone willing to listen: "Be careful what you wish for."

Dick's story is an excellent example of what it takes to be an exceptional leader. He understood that his first task as head of the IT department would significantly impact the organization and the people with whom he worked. Dick knew that at the same time that he was being promoted, others on the team would be losing their jobs. He also understood that this was necessary for the entire company, not just IT, and communicated that to his team. Dick courageously took responsibility for his decisions, handling layoffs himself rather than delegating them to staff members.

This story captures the highs and lows of being a leader, and the qualities necessary to succeed in a leadership role. Some of us accept these positions because we have reached the top of the technical career ladder, and it seems like the logical next step in our careers. Those who only see the opportunity to be a leader like that can quickly find the job daunting. Gone are the days when you were the technical expert and knew everything there was to know about your domain of expertise. When you become a technology leader, you must leave behind what is familiar and comfortable in exchange for what can sometimes feel overwhelming and frustrating.

Why abandon what is comfortable for something that can be so difficult? Dick Daniels exemplifies what is possible when you accept the challenges of leadership and embrace the possibilities it offers. Over his career, Dick led organizations that were recognized as one of the best places to work in IT for five straight years. During that same time, his team received the prestigious CIO 100 award for innovation five times and was eventually inducted into the CIO Hall of Fame. What Dick knew even before he received these honors was that a technology leadership role offered the opportunity to make a significant impact.

## The Privilege of Leadership: Making an Impact

This book will focus on the practices of technology leaders who have approached their role with an objective greater than merely pursuing the job title, status, or financial success. They embrace the ambiguity and stress that come with being a leader because they maintain their focus on why they wanted to be a leader in the first place. They have all taken on leadership roles to have an impact.

There are three critical areas of impact that these leaders have embraced in their role that will serve as the focus of this book:

- Impacting the firm through technology innovation
- Impacting the team by creating a great workplace
- Impacting the future of the technology profession

This book incorporates survey input from over 1,600 technology leaders in more than 60 countries regarding where they focus their efforts in these impact areas. While we all face different challenges as leaders in our profession, the survey results reveal how much we have in common despite where we live in the world, the industries we serve, or the size of our firms.

The leaders featured in this book have experience in various industries, including healthcare, financial services, manufacturing, government administration, and IT services. The fifteen leaders profiled have more than 500 years of combined experience in technology and have led organizations that received the CIO 100 award for innovation 33 times and Computerworld's "Best Places to Work in IT" 38 times. These leaders also acknowledge that their responsibilities extend far beyond the confines of their firm, and must include support for others on their journey into and through a career in the technology profession.

## Impacting the Firm through Technology Innovation

The company that hires you to lead a technology organization expects you to make a significant impact. This includes generating revenue, reducing costs, or gaining a competitive advantage. Impacting your firm will always be your priority as a leader and should never be dismissed as less important than any other impact area. If you do, you won't be in your role for long.

The stories the technology leaders shared for this book will emphasize their understanding of their employer's business, the firm's mission, and customers' needs. They have succeeded in positioning their organization as much more than a technology service provider. They recognize the full potential of technology to transform processes, products, and services, and have successfully established their organization as an innovation leader inside and outside their firm.

## Impacting the Team by Creating a Great Workplace

To drive innovation in an ongoing, sustainable manner, these leaders recognize that they must align their organization with the company's purpose, enabling each team member to understand their role in achieving the firm's mission. That brings us to the second impact area: your team. To impact your firm, you must have a highly engaged team that aligns with your vision and connects to the organization's mission. They must be resilient enough to weather economic downturns, changing priorities, and expanding business expectations. To have a team with these attributes, you must understand what motivates them both professionally and personally so that you can support them in their career journey.

The leaders featured in this book have led organizations that have received industry recognition as among the best places to work in the information technology profession. They recognize that fostering a great workplace for their team is crucial to achieving the company's objectives.

## Impacting the Future of the Profession

The final key impact area is the technology profession as a whole. To have a lasting impact on your company with an engaged team, it's essential to encourage and foster the development of the next generation of technology professionals. Leaders who excel in this impact area think many years ahead for the profession. They inspire high school students to pursue careers in computing by supporting science, technology, engineering, and math (STEM) programs in their community. They encourage university students to pursue computer science and information systems degrees through their commitment to mentorship and internship programs.

These leaders are also partnering with emerging alternative learning programs that focus on teaching technology skills to underrepresented populations, aiming to increase the much-needed diversity in the technology field. While these leaders may never benefit from hiring these individuals, they believe they are responsible for giving back to the profession and their community by offering their time and talents to develop the next generation of technology professionals.

Another way these leaders impact the profession is through their active involvement in organizations that support knowledge sharing and network building among technology professionals. They enable others in the computing field to gain insights from their successes and failures while recognizing the value of learning from others.

Finally, and importantly, they focus on the talent pipeline within their organization to ensure that they support the continued development of business, technical, and leadership skills that will be needed long after they depart from the firm.

## Three Qualities of Impactful Leaders

The leaders featured in this book demonstrate three qualities that enable them to make an impact on their firm, the team they work with, and the profession. They understand that before taking action to have an effect, they must first ***listen to learn*** about the goals, needs, and struggles of the people in each impact area. Once they gain this understanding, they can effectively ***communicate to connect*** with their audience. Ultimately, they recognize the need to ***act courageously and do the next right thing*** to achieve success and make a meaningful impact despite the risks of challenging the status quo.

## Listening to Learn Before You Speak

Listening can be challenging, especially for a leader managing the role's complexity and demands. Being an active listener means genuinely focusing on the speaker, not just appearing to pay attention while your mind wanders to other concerns like a system outage, a voicemail from the CEO, or a budget overrun. As a leader, listening can help you learn about issues you wouldn't otherwise understand.

In the following chapters, you will discover how influential leaders prioritize listening to their colleagues, teams, and hopeful technology professionals before expressing their views. They recognize that investing time to understand what these constituents face will yield valuable insights into how they can create the most significant impact. Whether in a business meeting, performance review, or mentoring session, being open to listening and learning will enhance your effectiveness as a leader.

## Communicating to Connect Your Audience to Your Mission and Vision

Connecting and communicating effectively is a fundamental quality displayed by the technology leaders highlighted in this book. They understand how to convey their message in different settings and can determine what, when, and how to communicate everything from the organization's strategy to difficult messages about project cancellations. They understand the significance of clear communication, knowing their audience carefully observes their words to ensure consistency with their actions.

The stories depicted in the following chapters use various communication methods that have inspired and motivated others. Critical message creation, delivery, and follow-up strategies will also be explored, along with techniques for utilizing credible sources to help convey your message.

## Doing the Next Right Thing to Have a Lasting Impact

Finally, this book will showcase leaders who consistently demonstrate the courage to take the right actions that create a lasting impact. Courage is a quality that separates these leaders from their peers. It has allowed them to present essential and sometimes controversial ideas for enhancing the firm's performance and the teams they lead. These leaders question conventional wisdom about the firm's operations and the policies governing their teams. They are not afraid to make the difficult decisions required for the company's success and the well-being of the individuals they lead.

This might involve recommending canceling a project that is unlikely to yield the expected benefits, challenging outdated company policies to provide employees with appropriate flexible and remote work options, or establishing an employee development program if team members' training needs are unmet. Having the courage to do the next right thing is crucial for making a meaningful and lasting impact on your company, the team you work with, and the technology profession.

In summary, the leaders featured in this book grasp that each impact area is interconnected and mutually dependent, as shown in **Figure 1-1**. They recognize the relationship between these three impact areas and understand that the qualities of listening, communicating, and doing the next right thing encompass all of them.

## Remembering My Purpose for Being a Technology Leader

Transitioning from a position where you excel as an individual contributor to one that requires you to lead others can be challenging. It's like coming to a crossroads in your career, where you have to choose between continuing with what you're familiar with or stepping into uncharted territory. Remember why you chose this path when you become a leader. Commit to memory what motivated you to make that decision, and periodically reflect on whether you have stayed committed to that purpose.

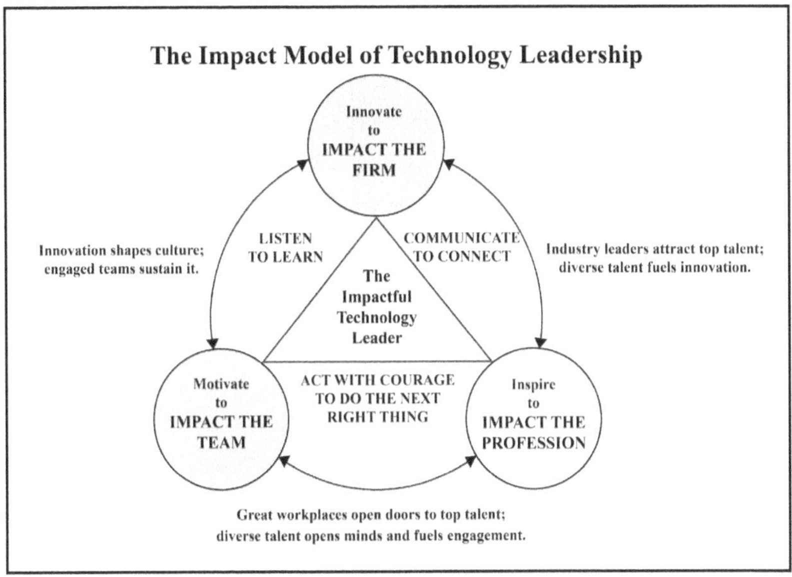


*Figure 1-1*

This is my story of rediscovering and embracing my purpose as a technology leader.

I knew I wanted to be a leader early in my career. In my late twenties, I had my first opportunity to head up a team of developers building software for nuclear power plants. That role allowed me to see how my decisions could enable the success of the company I worked for and each individual on the team. I was able to influence both technology and business decisions regarding the system we were building while also helping team members balance their career aspirations with their personal lives.

It became my purpose to be the best leader I could be. Following that assignment, I was fortunate to have supervisors who understood my career interests, recognized my potential, and acted on my behalf to help me become the leader I eventually became.

After a few decades in roles ranging from IT manager to IT director, I accumulated experience in multiple businesses, geographies, and increasingly larger teams and budgets. After nearly two decades at the company, I was offered the role of CIO. The current CIO, who had served for more than ten years, was preparing to retire, and the CEO reached out to share that I had been chosen as the successor for this position. I knew it was the only role I hadn't undertaken, and I accepted it as a challenge to see if I could succeed at the next level in my career.

After the first two years, I knew I could function as a CIO, but I began to doubt my ability to be good in the role. I had been successful in my previous positions, receiving positive performance reviews and knowing that my contribution made a difference. However, being a CIO left me wondering if I had accepted the role for the wrong reason. I soon had a chance encounter that helped remind me why I wanted to be a leader in the first place.

It was mid-February, and my wife, Janice, and I attended our church's Valentine's Day dinner. We were seated at a long table with other attendees, many of whom we didn't know. We sat across from a couple we hadn't met before, so we introduced ourselves and began discussing how we had heard about the event. During the conversation, I learned that the husband and I grew up in the same neighborhood and attended the same elementary school. It wasn't just any elementary school; it was the same school where my father taught fifth and sixth-grade classes.

Without him knowing that it was my father, I asked him if he had ever had Mr. Caruso as a teacher. He spent the next several minutes sharing with me what he had learned from my father as a 12-year-old boy and how it had remained with him throughout high school, college, and even into his 50s as an adult. From math to spelling and handwriting, my father's classes resonated with this man many years later.

Despite it being 40 years since he had my father as a sixth-grade teacher and more than 30 years since my father had passed away from cancer, Raymond Caruso was still remembered by one of his students as a passionate educator who truly cared about his students.

At that moment, I remembered why I decided to be a leader earlier in my career. It wasn't about accepting the next promotion, the compensation, or the title. I realized that I wanted to have the opportunity to positively impact the lives of others, just as my father did with so many of his students.

Before you move on to the next section, take a moment to reflect on your career so far. What were your personal and professional goals before you started that journey, and are you on a path that will help you achieve them? Do you know what you want to accomplish in your next position, and does that conflict with your personal goals? How can you learn about the expectations of a leadership position before accepting the role? Finally, and most importantly, reflect on your true purpose for wanting to become a technology leader and consider whether it holds more meaning than merely the next progression in your career.

# PART ONE

# IMPACTING THE FIRM: DRIVING INNOVATION THROUGH TECHNOLOGY

## CHAPTER 1

# Impacting the Firm: Why It Matters

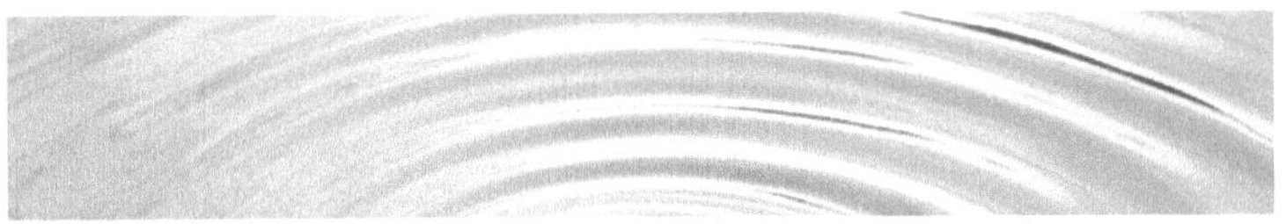

"Innovation distinguishes between a leader and a follower."
**—Steve Jobs**

Technology innovation enhances your firm's efficiency, transforms entire industries, and redefines the rules of competition. As a technology leader, you can take the initiative to drive change or allow others to dictate your impact in this role; the decision is ultimately yours.

Why don't all leaders in the field prioritize innovation? Often, they face competing priorities that prevent them from fully leveraging the potential of the technology for which they are responsible. These challenges include maintaining the firm's legacy systems and ongoing operations, customizing and upgrading applications and hardware, and defending against the latest cybersecurity attacks. While maintaining the firm's operations, they must monitor emerging technologies, recruit and develop talent to support these systems, and manage a flat budget.

The complexity that technology leaders are responsible for grows annually, with business leaders looking for every opportunity for operational efficiency and competitive advantage. These executives want the capabilities and benefits technology can offer without incurring additional costs.

Complexity also arises from the technology industry's business model, which is characterized by high switching costs and planned obsolescence, compelling continual upgrades, or the risk of increased maintenance costs or no support. For many leaders, this creates a "technology treadmill" where they run as fast as possible after the next "new thing," all while trying to maintain their current state, only to feel that they are not making progress.

Another major factor impacting the technology organization's ability to pursue innovation opportunities is the ever-changing issues facing the business it supports. Whether facing economic downturns, competitive pressures, acquisitions, divestitures, or changes in senior leadership, your agenda can change rapidly. Resources can be shifted between projects as business priorities change, causing some leaders to think, "Why bother?" when adding innovation to their agenda.

The leaders profiled in Part 1 understand the realities of directing a technology function and don't allow these to be distractions that deter them from their mission to have an impact. They view innovation as essential to their organization's value, like system availability, data security, and business continuity. Innovation is frequently integrated into their organization's daily operations. At the start of a new project, time and tasks are planned to help the team discover innovative approaches to developing a solution.

When business challenges emerge, these leaders don't wait for someone to come forward with a technology or system request. Instead, they take the initiative to proactively organize their teams and develop solutions to address these issues, regardless of whether technology is involved in the solution or not. They don't need expensive tools, high-tech labs, or external consultants to enable meaningful innovation. They leverage their existing budget and talent on their teams. In summary, these leaders view their innovation practice as integral to their organization's identity and culture.

## Getting Off the Technology Treadmill and Becoming an Innovation Driver

Just because CIOs want to spend more time on IT-led innovation doesn't mean that their teams are prepared for this challenge. Some view technology professionals as being "left-brained"; logical, analytical, and problem-solving. It has often been the belief that only "right-brained" people who are creative and imaginative can be innovative. That myth is dispelled in *The Innovator's DNA: Mastering the Five Skills of Disruptive Innovators by Jeffrey H. Dyer, Hal B. Gregersen, and Clayton M. Christensen. Harvard Business Review Press, 2019.* In this book, the authors note that "creativity is not just a genetic predisposition; it is an active endeavor" and that innovative behaviors can be learned. The authors conducted research into the practices of highly successful innovators, including Jeff Bezos of Amazon, Scott Cook of Intuit, and Michael Dell of Dell Computer. Their research resulted in the identification of the behaviors that these innovative leaders have in common. They refer to these behaviors as the five "discovery skills" of experimenting, questioning, observing, networking, and associating.

- **Experimenting** often comes to mind when we think about being innovative. The leaders that the authors cite are apt to try new experiences, deconstruct products and processes, and develop quick, inexpensive prototypes.
- **Questioning** is a critical skill in unlocking new ideas by challenging the status quo to understand the current state and what might be possible. Questioning, when done well, can make others feel uncomfortable. It is meant to test norms and "the way we have always done things" and when conducted effectively, can uncover new opportunities to improve products, processes, and services.
- **Observing** requires the innovator to go to where their customers are trying to get their work done. Whether it is an internal or external customer, it is important to understand the job that they are trying to complete and the challenges that they have in getting it done. It is through observation that insight can be gained into the customer's struggle and the potential solutions to address those challenges.
- **Networking** for ideas occurs when insights are solicited from individuals with diverse backgrounds and perspectives. These can include people from a different industry, generation, gender, or geography. Networking can also involve traveling to other parts of the world and attending conferences outside of your field or industry.
- **Associating** involves taking the output from experimenting, questioning, observing, and networking and creating a connection with something from a completely different application. It is a cognitive activity that occurs when you connect ideas from an industry, profession, or geography outside of your own to create an innovation to address a problem that you are trying to solve.

Technology organizations can leverage all five discovery skills to drive innovation. The following examples, based on real situations the author has encountered, demonstrate how each skill works in practice.

## Experimenting

Experimentation does not require large budgets or dedicated teams. Instead, it requires that you regularly monitor emerging technologies, understand how your competitors may be employing them, and consider how your company could benefit from adopting a new capability made possible by them.

One powerful approach to low-cost experimentation involves conducting regular emerging technology reviews with your team and business colleagues. These sessions position you as a forward-thinking leader while keeping your organization ahead of technological shifts that could impact your industry. The format is straightforward: select an emerging technology, whether artificial intelligence, blockchain, edge computing, or another relevant innovation, and facilitate a collaborative discussion of what it is, how it works, and where it might create value for your business.

Using accessible tools like PowerPoint, Figma, or Miro, you can rapidly prototype visual demonstrations of how the technology could be deployed within your firm, illustrating potential workflows, user experiences, or process improvements. This approach delivers dual benefits: it educates business stakeholders about technologies they may otherwise dismiss as too complex or abstract while collaborating on innovative application ideas from people who deeply understand your organization. The cross-functional conversation often reveals unexpected use cases that neither technologists nor business leaders would have identified working in isolation.

Another valuable experimentation approach involves systematically analyzing technology-based competitor patents to decode their strategic intentions. This reverse engineering process goes beyond simply reading patent claims; it requires deconstructing the patent to understand the underlying business problem the competitor identified, the technical approach they've chosen to solve it, and the competitive advantages they hope to gain. By working through this analysis with your team and business partners, you can increase your strategic intelligence about where your market is heading.

This patent review methodology delivers multiple strategic benefits. First, it demonstrates to business stakeholders that your technology organization actively monitors the competitive landscape and understands threats before they materialize in the market. Second, this analysis reveals gaps and opportunities in your own innovation pipeline. You may discover that competitors are pursuing approaches that validate your current direction, suggesting you should accelerate development. Alternatively, you may identify entirely different solutions that are more cost-effective or aligned with your organization's capabilities. Finally, understanding the specific claims in competitor patents helps your team innovate around them, developing differentiated approaches that deliver similar benefits without infringement while potentially offering superior customer value.

## Questioning

Questioning is critical in unlocking new ideas by challenging the status quo to understand the current state and what might be possible. When done well, questioning can make others feel uncomfortable. It tests norms and "the way we have always done things." When conducted effectively, it can reveal new and transformative ideas.

Asking questions is an inherent part of what many in technology do daily. If there is a request for a new system to be built, we ask questions regarding what it should do and how it should integrate with other systems. We ask who the users are, how many will use the system, and whether they require remote access. We ask many questions, but they are usually directed at understanding what the user wants. We often avoid difficult questions such as, *"Why do you want to automate this?" "What do our competitors do to address this?"* or *"How will this enhance your efficiency or increase the company's revenue?"*

The ability to ask probing questions transforms the existing practices of many technology organizations into something more meaningful. It prompts the person being questioned to think critically about their needs and why a technology solution is the most appropriate way to address them. Developing the ability to ask insightful questions helps uncover the real problem your customer needs to solve and opens up possibilities for innovative solutions.

One powerful approach to developing the skill of questioning is to examine situations from perspectives other than your own. For example, imagine how a customer might view the problem you are trying to solve: What questions would they ask about making your website more engaging or doing business with your company easier? Consider what your board of directors may ask about the investment you are proposing, or anticipate the concerns employees would raise about implementing a new system and its accompanying processes. In each of these scenarios, the questions your constituents ask, whether customers, board members, or employees, will likely differ significantly from yours and tend to be more critical, revealing blind spots in your thinking.

Another approach that proves especially effective for technology organizations is to explore how emerging technologies could transform your firm's future. Ask insightful questions about your industry's direction, and how new technologies might influence it over the next three, five—or even ten—years. This forward-looking inquiry allows you to proactively shape your organization's trajectory rather than scrambling to respond after competitors have already moved ahead. By consistently questioning what's next, you position your firm to lead change rather than follow it.

## Observing

Observing requires the innovator to go where customers try to accomplish their work. Whether it's an internal or external customer, understanding the job they need to complete and the challenges they face in achieving it is crucial. Through observation, insight can be gained into the customer's struggle and the potential solutions to address those challenges.

Observation almost always requires leaving your office, cubicle, or home and traveling to where your customer is located. If your customer is another employee at your work location or a remote site, go to where they are working and watch what they do. Ask them where they face the most difficulty in completing their daily tasks and learn about the workarounds that they have created to get their jobs done. The "unsanctioned" methods and tools they have created often provide insight into innovation opportunities.

If your customer buys your product in a retail store, take the time to understand how they decide between your product and that of a competitor. Investing time observing, listening to, and understanding your customers' purchasing challenges can enable you to identify new and improved ways to assist them in completing their purchase.

## Networking

Networking for ideas occurs when insights are collected from individuals with diverse backgrounds and perspectives. These may include people from different industries, age groups, genders, or geographic regions. Networking can also involve traveling to other areas or attending conferences outside your geography, field, or industry.

Local or national professional organizations and conferences can provide a rich source of innovative ideas. Meeting new people and learning about their industry, customers, and how technology is used to solve their business problems can inspire your creativity. By attending these events and networking for ideas, you may discover new and improved ways to add value to your firm.

As a technology leader, one of your most valuable resources is your network of professional peers. Whether through local user groups, industry associations, or national organizations, cultivating a supportive community of technology professionals provides access to diverse perspectives and proven solutions for current challenges and future opportunities. These relationships create a safe space to discuss sensitive issues, share lessons learned, and explore emerging trends without the competitive pressures within your organization.

The tangible benefits of such networks can be substantial. Peer connections have yielded low-cost, creative approaches to addressing cybersecurity threats that smaller organizations might otherwise struggle to afford. They've revealed effective marketing strategies for internally marketing the technology function, helping CIOs gain executive buy-in for critical initiatives.

Network members frequently share supplier negotiation tactics and vendor performance insights that can save significant time and money. Beyond solving immediate problems, these relationships often spark innovative ideas through exposure to how peers in different industries approach similar challenges.

Another powerful way to leverage networking for innovation is through mentorship of emerging talent, whether inside or outside your organization. The mentor-mentee relationship is bidirectional. As a mentor, you often gain as much insight as you provide. Younger professionals bring fresh perspectives on emerging technologies, different problem-solving approaches, and valuable insights into generational differences that affect employee motivation and engagement. They challenge assumptions you may not realize you're making and expose you to cultural shifts that might otherwise remain outside your awareness.

## Associating

Associating involves taking the output from experimenting, questioning, observing, or networking and creating a connection with something from a completely different application. This cognitive skill lets you link concepts from disparate industries, professions, or geographic contexts to develop innovative solutions. Identifying the connection between two or more unrelated things requires imagination and practice; however, it can ignite your team's creative capabilities when executed effectively.

Consider the following example from the 2010s: when toothbrush manufacturers introduced "smart" electric toothbrushes that tracked brushing frequency, duration, and technique, they solved a problem about improving consumer health habits through real-time feedback. A skilled observer may associate this innovation with a different domain: applying paint to surfaces in residential or commercial buildings. Paintbrush and paint roller manufacturers could incorporate similar sensor technology to help professional painters monitor whether they're applying paint uniformly and efficiently across large surfaces, potentially reducing material waste and callbacks for touch-up work.

The same technology could transform the experience for do-it-yourself painters, who often struggle with technique. Smart rollers could deliver haptic or visual feedback when users press too hard, preventing streaks and roller marks that frequently ruin painting projects. While both applications involve brushes, one delivers toothpaste to teeth while the other applies paint to walls; they are seemingly unrelated tasks. Yet this example illustrates how a solution successfully implemented in one industry can inspire innovative approaches for different applications.

## Which Innovation Skills Are Practiced by Technology Leaders?

While understanding these five discovery skills is useful, it's even more important to know which skills technology leaders are using. To better understand how these skills are utilized across the broader technology leadership community, over 740 senior technology leaders from 47 countries responded to a survey asking: *Which of the five discovery skills have you fostered in the organizations you've led to build a culture of innovation?* The responses were analyzed across multiple dimensions, including country, gender, industry, and leadership level. The results are detailed in Appendix A.

**Figure 3-1** shows the responses from all survey participants. The most common discovery skill is questioning, with an 89% response rate, while the least common is associating, at 63%.

| Which of the discovery skills have you fostered in the organizations you have led to create a culture of innovation? | | |
|---|---|---|
| **Discovery Skill** | **% of All Respondents** | **Rank** |
| Experimenting | 78% | 2 |
| Questioning | 89% | 1 |
| Observing | 77% | 3 |
| Networking | 75% | 4 |
| Associating | 63% | 5 |

Figure 3-1

By segmenting the results by position held by the respondent, some differences begin to emerge. **Figure 3-2** shows the feedback from senior executives with CxO titles, including Chief Information Officer (CIO), Chief Technology Officer (CTO), and Chief Digital Officer (CDO), as well as those with titles of Vice President of Information Technology, and Head of Information Technology. This feedback is compared to the input from respondents with Chief Information Security Officer (CISO) and Director/Manager titles.

| Which of the discovery skills have you fostered in the organizations you have led to create a culture of innovation? | | | |
|---|---|---|---|
| **Discovery Skill** | **Senior Executives** | **CISO** | **Director / Manager** |
| Experimenting | 80% | 71% | 61% |
| Questioning | 89% | 89% | 87% |
| Observing | 78% | 71% | 65% |
| Networking | 75% | 85% | 68% |
| Associating | 63% | 62% | 58% |

Figure 3-2

These findings reveal distinct differences in innovation behaviors based on organizational role and seniority. Senior leaders are likelier to encourage experimentation (+19%) and observation (+13%) than their Director or Manager counterparts. Interestingly, Directors and Managers report the lowest engagement in fostering the critical skill of associating. This gap may reflect a lack of empowerment at the middle management level to drive innovation initiatives. Yet these leaders are often closest to day-to-day operations and best positioned to influence innovation outcomes across their teams.

Those with senior executive titles differed from CISOs in their responses regarding experimenting (+9%) and observing (+7%). However, they were less likely to foster networking (-10%) than the CISOs responding to this survey. One reason for this difference in networking might be that many CISOs rely on organizations that share information and strategies to tackle cybersecurity issues. For many, networking for ideas is crucial for CISOs to be successful.

**Additional high-level findings from the survey include:**
- Geographic consistency: U.S. and non-U.S. respondents reported similar levels of focus on innovation skills, suggesting a consistent approach across regions.
- Gender insight: Female senior leaders are 9% more likely than males to network for new ideas.
- Company size impact: Senior leaders at organizations with more than 10,000 employees are 10% more likely to foster experimentation than those at smaller firms.

## Results from Innovative Technology Leaders

The data was further segmented to identify respondents whose organizations had received the CIO 100 Award to gain deeper insight into the practices of leaders recognized for innovation. Fifty-four survey participants led organizations honored with this prestigious recognition one or more times. Presented annually by Foundry's CIO, the CIO 100 Award celebrates organizations that drive business value through innovative digital solutions. These award-winning solutions provide a competitive advantage while improving processes and customer relationships.

**The following is a summary of the responses from CIO 100 leaders participating in this survey:**

- A total of twenty-eight industries are represented, including healthcare, government, financial services, and IT services.
- While 80% of respondents are from the U.S., the other countries represented included leaders from India, Cyprus, and Sweden.
- The survey participants are 81% male and 19% female.
- 48% of the participants come from firms with more than 5,000 employees, while 52% of those responding are from firms with fewer than 5,000 employees.

## What Do Leaders Recognized for Innovation Do Differently?

**Figure 3-3** illustrates the data from leaders receiving the CIO 100 award and their senior executive counterparts holding CxO, VP IT, or Head of IT titles.

| Which of the discovery skills have you fostered in the organizations you have led to create a culture of innovation? | | | |
|---|---|---|---|
| Discovery Skill | CIO 100 | Senior Executives (Excluding CIO 100) | Difference |
| Experimenting | 93% | 78% | +15% |
| Questioning | 94% | 89% | +5% |
| Observing | 83% | 77% | +6% |
| Networking | 74% | 75% | -1% |
| Associating | 74% | 63% | +11% |

**Figure 3-3**

Several notable differences emerge when comparing the results of these two peer groups. Specifically, leaders in CIO 100 recipient organizations indicated they foster four of these five skills more than their peers. They place significantly greater emphasis on experimenting (+15%) and associating (+11%) than their peers. By focusing more on experimentation, they invest greater effort in assessing the viability of new ideas through small pilot projects and prototypes than their counterparts. They also pay significantly more attention to the cognitive skill of association by using the results from practicing other discovery skills to identify solutions that can address their needs, which may have originated in a completely different industry or part of the world.

Notably, 50% of CIO 100 leaders surveyed reported fostering all five discovery skills, compared to 37% of other respondents. Whether or not they've read *The Innovator's DNA: Mastering the Five Skills of Disruptive Innovators*, these leaders have embraced its core practices within their organizations. The fifty-four CIO

100 leaders have collectively led organizations recognized with the award 75 times. They understand that innovation doesn't happen by chance; it must be intentionally nurtured across the enterprise. In many ways, these leaders assume the role of Chief Innovation Officer, whether formally or informally, fostering a culture that encourages innovation to thrive.

## CHAPTER 2

# Impacting the Firm: How Do They Do It?

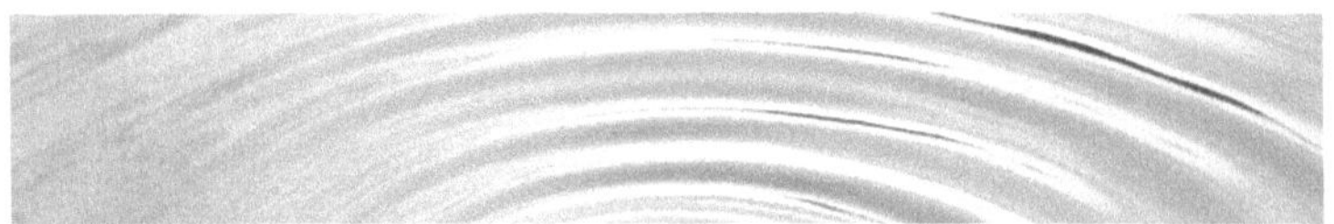

## VISHAL GUPTA

Discovery Skill: **Experimenting**

Firm: **Lexmark**

Vishal Gupta served as the Chief Information and Technology Officer (CITO) at Lexmark, a leading provider of printing and imaging technologies and supporting software and services. Gupta is a leader who understands that a culture of employee-driven innovation is critical to the company's ongoing success. The company's mission captures the essence of this culture of innovation at Lexmark:

> "We work with passion, conviction, and speed to solve challenges, create new opportunities, and enable responsible operations for partners and customers. We power this future through innovative people, technology, and services."

Gupta and Lexmark's leadership recognize that the innovation of their products and services depends on the culture they establish for their employees. According to Gupta, three key elements are required to build and maintain this innovative culture. First, all aspects of the business strategy should be powered and supported by innovation. Secondly, the firm's structure must support the enablement of radical forms of innovation. He notes that once the proper structure is in place, innovation isn't hindered by organizational thinking that says, "We've never done this type of thing before." And finally, the firm needs to "make space for innovation and support it as an organization."

Lexmark makes "space for innovation" through institutionalized employee-led experimenting called "Focus to Future," or F2F. On the surface, F2F appears to be a classic idea hackathon approach for generating innovative ideas. Lexmark, however, has modified the F2F format over several years to serve as a source for identifying breakthrough approaches that address the firm's needs. These events occur twice a year and take place at multiple locations worldwide. Teams of two employees are assigned to tackle high-level business challenges over three days, focusing on areas ranging from business operations to new product and service features.

Gupta points out that Lexmark experimented with different F2F formats and concluded that, since innovation is a "team sport," the structure of each team should be carefully considered. The leadership team saw the F2F event as an opportunity to promote collaboration and identify innovative ideas generated by their employees. To achieve this, each team consisted of two employees who didn't work directly together daily. This enabled them to ensure that different perspectives would be represented on each team, and as a result, they were able to develop more breakthrough ideas. The company also views these events as an opportunity to develop the communication skills of its employees. Therefore, each team must create a video of their idea and a high-level prototype that explains their concept in detail.

As the F2F format evolved, Lexmark aligned the event with the product management process to help the company address some of its most pressing needs. As a result, employees know their efforts will contribute to the company's future success and will not be forgotten after the event concludes.

Lexmark also reinforces its support by rewarding the best-proposed innovations and building excitement around the F2F events. Teams are showcased in a symposium-like environment where attendees include employees and external guests. As a result, participating teams gain recognition for their efforts and have the opportunity to receive prize money to develop their prototypes further.

A similar innovation competition organized by Lexmark resulted from connecting employees with the challenges faced by the company's end consumers. For many healthcare professionals, providing information for follow-up appointments or how to care for a patient may not be available in the language native to the patient or their caregiver. Additionally, in a school setting, a teacher may need to provide homework information to a student whose parent or guardian speaks a different language. This resulted in an innovative challenge of identifying a way to translate a document from one language to another during printing. The result was the development of Translation Assistant, a real-time document translation service available via a Lexmark multi-function printer.

Translation Assistant enables users to scan a document and select the original and target languages for translation. The document is then transmitted to a cloud service, where it is transformed into one of more than two dozen languages and formatted. After completing this, it is sent back to the printer, where it can be printed or emailed. The result is a solution for Lexmark's customers that enables them to communicate more efficiently and effectively.

Another idea generated from an F2F event was focused on an internal opportunity to improve operations within their factories. The winning idea comprised a camera that would visually inspect activities in the factory and, using artificial intelligence, identify quality issues in the manufacturing process. The idea was fully developed and implemented, enabling Lexmark to proactively predict 95% of problems before they occur at the factory and remotely resolve 70% of the issues. While initially focused on their internal operations, the company has since expanded the concept to other devices and commercialized it as the Lexmark Optra Edge.

By cultivating a culture of innovation through structured, company-sponsored experimentation, Lexmark has created a scalable process for engaging its workforce in solving significant business and customer problems. Rather than limiting innovation to R&D departments, the company has embedded it across the organization, encouraging employees at all levels to contribute ideas, test new concepts, and collaborate across traditional boundaries. This inclusive and disciplined approach generates a continuous stream of relevant innovations, deepens employee engagement, reinforces a shared sense of purpose, and fosters a connection between employees and the company's mission.

## Practicing the Skill of Experimenting:

Here are some methods that you can try to gain new perspectives and insights.

- **Map Customer Journeys:** Analyze internal and external customer processes to identify pain points, inefficiencies, and opportunities for improvement. Use this insight to rethink how value is delivered.

- **Host Hackathons with External Participants:** Invite interns, customers, and suppliers to participate in idea hackathons to gain fresh viewpoints on how to solve a challenge your firm is facing.

- **Collaborate with College and University Students:** Engage with colleges and universities that offer capstone courses for their students to apply the skills they've learned, providing your firm with new ideas to solve business problems.

- **Form Tech and University Knowledge Sharing Alliances:** Build technology exchange partnerships with academic institutions or innovation-focused companies. Gaining access to early-stage research or emerging technologies can inspire new avenues for experimentation.

- **Benchmark Your Competition:** Compare your online purchasing experience with key competitors. Evaluate usability, speed, and customer friction points to uncover areas where your process can advance significantly.

# FORTUNE 200 CIO

Discovery Skill: **Questioning**

Firm: **Fortune 200 company**

Asking "who, what, where, when, and why" is a time-tested approach in designing systems, analyzing workflows, and driving process improvements. In the technology world, these questions are often applied with a solution already in mind, used to confirm assumptions or fine-tune execution. But what if we redirected our questioning away from the solution and instead toward the heart of the problem? What new insights might emerge if we challenged the underlying assumptions before rushing to fix them? This shift from solution-driven to problem-driven inquiry is at the core of innovative leadership. The following story explores how one award-winning CIO adopted this mindset to foster deeper understanding, spark creative thinking, and deliver breakthrough results.

The CIO of this Fortune 200 manufacturer observes, "I've always found it challenging when you're in an environment where you have a strict deliverable, a very tight budget, and constrained resources and timelines, making it difficult to help your team feel like they have the opportunity to be innovative." Many IT organizations view themselves as delivery or service organizations rather than innovators. In this situation, these organizations view themselves as order-takers. This CIO supported her team in developing their innovation skills, enabling them to think about business problems from the perspective of how technology could be applied in innovative ways to find solutions.

One example of how this leader focused her team on solving business problems involves the discovery skill of questioning. She reflects that "One of the problems that I see in many IT organizations is that instead of investigating the problem, they start by jumping to technology solutions. The first thing they want to do is discuss the technology and the end solution. I feel like you often miss the opportunity to be innovative when you jump to the solution too quickly and don't focus on peeling back the onion around the problem." Her approach to gaining deeper insight into the problem involved asking "why" questions. This CIO learned this behavior from some of the more innovative leaders she had worked with throughout her career. These leaders posed probing questions to thoroughly understand the issue before proposing a solution.

The following is one approach that this leader used to develop her team's questioning skills. Although the exact business issue and outcome from this strategy session must remain confidential, the approach taken by this CIO illustrates an effective method for enhancing questioning skills within your team.

This CIO's method for enabling a structured questioning approach among her team was to conduct what was referred to as a "strategy session." This typically took place over ninety minutes, during which a problem facing the business would be presented. In one instance, the session focused on an issue that the senior leaders in the company were grappling with and didn't have a known solution for. After detailing the problem, she told her team, "Today, we are not discussing solutions; we are discussing problems." For the remainder of the time during that initial strategy session, the team was encouraged to ask any questions to improve their understanding of the issue. Although the details of the challenge must stay confidential, the questioning process illustrates the effectiveness of this method. Rather than "What technology could resolve this?" the team discovered that they should start by asking, "Why does this problem exist?"

At the end of the first meeting, the team was informed that a follow-up session would occur in a few weeks and would be assigned a task to consider. They were asked to imagine themselves as a group of competitors working in a garage with knowledge of this problem and a single goal: to disrupt the company's business. Their role in this assignment was not that of someone in a technology organization but as a businessperson with the singular goal of putting the company out of business. Their assignment would be to develop business ideas that capitalize on the issue the firm's senior leaders were facing.

In a few weeks, the team regrouped for another ninety-minute session, and the CIO told the attendees that "the only rule was that there weren't any rules." There were no boundaries put on what could be covered. The group didn't need to identify technological solutions. The team could discuss potential business process changes or any other solutions. They believed that approaching the problem from the perspective of the competitor in the garage who wants to put the company out of business enabled the team to be free from any constraints in identifying solutions. The CIO noted that the team developed "many ideas during this session, some of which might have been considered crazy. The crazier, the better because I wanted the team to stop feeling like they had boundaries on how to approach a problem or a solution." As the session progressed, "People got excited and started jumping up and drawing pictures on whiteboards, getting so excited that they would lose their inhibitions and explore."

The second strategy session resulted in several ideas that the group considered viable solutions to the company's problem. At that point, the CIO informed the group that she was stepping out of the process and was no longer leading the sessions. The team would need to determine what to do next and how to organize and explore these ideas further. This approach aimed to identify the natural leaders and risk-takers within the group and determine who possessed leadership qualities and felt empowered to take on this challenge.

The CIO committed to presenting the team's proposed solutions to senior leadership within the firm. Several of the innovative ideas proposed by the team were eventually approved for implementation by the company. More importantly, the IT function was recognized as being proactive business partners who promoted new solutions rather than "order takers" waiting for the business to come to them with a problem to fix.

The company's CIO had extensive experience leading teams and driving innovative efforts. She understood that the first step in identifying ideas to solve a seemingly insurmountable problem is to ask as many questions as possible. By focusing on asking questions rather than seeking solutions, this team gained a deeper understanding of the business issue and broadened their options for identifying innovative solutions.

## Practicing the Skill of Questioning

This story demonstrates how questioning can be a starting point in the innovation process. It challenges conventional wisdom, uncovers root causes, clarifies objectives, and promotes better decisions. Building on this CIO's approach, here are some practical questions organized by everyday business situations that serve as examples to help teams develop stronger questioning skills and become more confident in using questioning in their daily activities.

## Questions to Consider Asking Before Starting a New Project

When launching a new initiative, thoughtful and challenging questions are essential to ensure the investment is appropriate and aligned with the organization's priorities.

- Why is this investment necessary, and what will happen if we don't pursue this project right now?
- Why will this benefit our customers when it is implemented?
- Why is this more beneficial to the firm and a better investment than other potential projects?
- Are we currently managing too many other initiatives?
- Can the organization support the change this will introduce?
- Will this change affect existing business processes? Has the process been reviewed and optimized for the new solution and revised business method?
- What are our competitors doing to address this need?
- If our competitors have this capability, do we understand the benefits it provides them?
- How can we mitigate the risks associated with the project?
- At what point would we terminate the project if the benefits appear unachievable?

## Questions to Consider Following a System Failure or Service Disruption

Whether responding to a failed project, a cybersecurity incident, or a system outage, a post-mortem review is a crucial opportunity to ask tough questions, gain insights, and enhance resilience.

- Why did this happen?
- Do we understand the root cause of the problem?

- What is the impact on our business, customers, employees, and community?
- How did we inform our constituents about the issue, and were we prompt in doing so?
- Did we foresee this as a possible occurrence, and if so, what was our contingency plan?
- If we had a contingency plan, was it effective, or do we need to revise it?
- Was the contingency plan tested under real or simulated conditions to ensure it would work?
- Did our technology suppliers respond promptly to assist us in addressing this?
- Have other firms faced this issue, and how was it addressed?
- What downstream impact does this have on other systems and related processes?

## Questions to Consider
## When a Team Member Leaves

If someone decides to leave your team and an opening arises, consider asking some of the following questions to better understand their reasons for leaving and how to fill the role effectively.

- Do we know why this employee left?
- Did we try to retain them? If we did, why was our offer declined?
- Was their departure due to a performance issue?
- Did the former employee indicate that their supervisor was the reason for leaving? If so, have other former employees also cited this supervisor as their reason for quitting?
- Do we need to fill the position, or can the work be eliminated?
- If we can't eliminate the work, can it be redistributed to other team members?

- Is there an internal candidate who could benefit from the career development opportunity this role would provide?
- Could another employee expand their current scope and take on these responsibilities?
- Does this role have a backup? If not, why?
- Would this position be better served by a contract or a temporary employee?

The power of questioning is not only in finding answers but also in asking better questions. As this CIO demonstrated, shifting from "How do we implement this solution?" to "Why does this problem exist?" can transform a team from order-takers to innovation partners. In a world where technology solutions are plentiful, the competitive advantage belongs to those who master the art of asking the right questions first.

## CHRIS CARUSO

Discovery Skill: **Observing**

Firm: **PPG**

As technology professionals, we rarely find the answers we seek by only asking questions. They can hide in plain sight; in the hesitation before a user clicks a button, in the workaround an employee has perfected, or in the pattern of data that doesn't quite fit the model we expected. Our ability to observe often decides whether we create solutions that genuinely make a difference or merely fulfill the minimum requirements.

Observation takes many forms in our work. It might mean shadowing a worker through their daily routine to understand the gap between how a process was designed and how it's actually performed. It could involve watching customers interact with your product, noting not just what they do, but what they struggle with, avoid, or improvise. Sometimes it means stepping back to study how an entire system operates, where bottlenecks form, and where the unexpected happens.

Observation can also include visiting a facility in person to understand the complexities that no requirements document captures: the impossible distance between workstations, the surprising demographics of the workforce, and the actual conditions under which your software must perform.

It extends into the digital realm as well. Monitoring technology trends helps you distinguish genuine innovation from fleeting hype. Analyzing website analytics, what content draws attention, how long engagement lasts, and what paths users take through your site reveals the story of how people actually experience your digital presence.

The following examples, each drawn from my own experience, demonstrate how the discovery skill of observation appears in a variety of situations and provides insights that change our understanding of the problems we're trying to solve.

In the do-it-yourself consumer paint business, a colleague spent time observing customers at a home-improvement center, immersing himself in the real-world challenges shoppers face. During one such visit, he saw something that would fundamentally change his understanding of customer behavior: a shopper carrying a bathroom rug through the aisles, eventually reaching the paint department. There, she carefully held the rug up against paint chip after paint chip, comparing colors until she found the perfect match to complement the rug she wanted to buy. What might have seemed like a one-time event was actually a glimpse into a common customer problem: the struggle to coordinate colors across different home décor categories.

This simple but revealing observation sparked a powerful insight: customers weren't just buying paint in isolation; they had a decorating project that they wanted to complete. The woman with the rug was solving a difficult color coordination problem through trial and error, physically transporting items across the store to make visual comparisons. She represented countless customers who lacked confidence in their color-matching abilities or who wanted assurance that their purchases would look good together. The challenge wasn't just about finding the right shade of blue; it was about understanding how that blue would interact with rugs, curtains, furniture, and accessories scattered throughout different departments in the store.

Recognizing this opportunity, a method was developed and patented to create a repository of home décor items linked by color. The resulting system would allow consumers to take any home décor item to a kiosk in the paint department, scan the item's barcode, and immediately see coordinating paint colors along with other home décor in the store that would complement the original item. This would enable the retailer to use the kiosk for cross-selling a wide range of merchandise that the customer could incorporate into their project. It would also give the paint manufacturer the chance to sell paint to customers who might not have initially considered painting their room as part of their decorating plans.

This same capability could be integrated into a retailer's website. Like many online retailers, when a customer views an item, such as curtains, they are often shown other curtains they may want to purchase. With the home décor coordination system, the retailer could offer the online shopper a wide selection of "certified" color-coordinated items, including paint colors. This would create an opportunity to ease the often challenging decision-making process of choosing purchases based on the item's digital color representation. By pre-matching the items and storing them in a database, the customer could trust that their project-related purchases would be color-coordinated.

What began as a single observed moment, a woman holding a rug against paint chips, evolved into a comprehensive solution that reimagined how customers approach home decorating, transforming a fragmented, anxiety-inducing process into a seamless, confidence-building experience.

Another powerful way to observe customer behavior is by analyzing website clickstream data. This analysis reveals which content is most important to customers and which information is less valuable, leading to improvements in website layout, content priorities, and unexpected strategic insights. By tracking user navigation patterns, time spent on pages, and engagement with specific features, companies can gain a detailed understanding of customer interests and decision-making processes. This data-driven approach often uncovers surprising patterns that challenge conventional assumptions about customer preferences and behavior, allowing businesses to make decisions based on data rather than just instinct.

A compelling example emerged from analyzing traffic on a website featuring paint colors matched to professional sports teams, allowing fans to paint rooms in their favorite team's colors. The assumption was initially that fan interest would remain relatively steady throughout the year, with perhaps modest increases during the regular season. However, the clickstream data told a dramatically different story about when and how customers engaged with the site.

The data revealed a striking pattern: interest in specific team colors surged dramatically during playoff periods across all sports. For example, during the two weeks leading up to the Super Bowl, the two competing teams' colors received more views than all other teams combined for the entire season. Similar spikes occurred during the NBA Finals, World Series, and Stanley Cup playoffs, with the intensity of interest correlating directly to how far teams advanced in their respective tournaments. This pattern held across all major sports, demonstrating that emotional investment in team success directly translated to purchase consideration.

This insight uncovered new ways to significantly increase online paint sales during peak fan engagement times through targeted marketing and promotional timing. The company could now launch just-in-time advertising campaigns focused on advancing teams, create special promotions, and optimize inventory based on which teams were still competing.

Both examples reveal a fundamental gap between how companies assume customers think about their products and how customers actually behave. Paint isn't simply a wall covering; it's an element in a larger decorating project that customers struggle to coordinate. Sports team colors aren't a year-round consideration; they become an impulse-purchase driver during the intensity of playoff competition. These emerged only through careful observation: watching a woman carry a rug through store aisles and analyzing clickstream patterns during championship runs.

The power of observation lies in its ability to reveal the hidden logic of customer behavior, the unspoken needs, the unexpected timing, and the workarounds people develop when products don't fully meet their goals. Whether shadowing customers in physical spaces or analyzing data from digital interactions, observation reveals not just what people do but also why existing solutions fall short and where true innovation opportunities exist. When we move beyond relying solely on what customers tell us and start paying attention to what they demonstrate through their actions, we shift from order-takers fulfilling requirements to problem-solvers creating solutions that customers didn't even realize they needed.

### Practicing the Skill of Observing:

Observing can take many forms. Here are a few additional examples to consider when developing the ability to master observation.

#### Observing Internal Customers

- Visit your customers' workplaces to observe how they perform their jobs. Hans Keller, CIO of Erickson Senior Living, notes, "You can't innovate from your cube. The only way to know what needs to be done is to go out and spend a day wandering around."
- Examples of "wandering around" include traveling with a salesperson when they visit their customers, working at a service or retail location, a warehouse, or a manufacturing site, listening to customer and technical service calls, and shadowing other departments, such as human resources, finance, and procurement.

#### Observing External Customers

- If your firm goes to market through a retail store, visit the store to observe customers' purchasing decisions. Do they purchase your product in conjunction with another? If so, what else are they buying, and how do they use it alongside yours?
- Visit a customer or a customer's customer to observe how they utilize your firm's product or service. Inquire about

what they like or dislike regarding the product and their experience with your firm. Are there any pain points in doing business with your company? If so, what changes would your customers suggest to enhance this experience?

### Observing Competitors

- If you work in an industry where you sell products directly to consumers or through distributors, visit competitors' stores and distributors to observe how they merchandise, sell, and support their products.
- Conduct and present a competitive analysis of your competitors' technological capabilities to determine whether they gain an advantage from using those technologies. Compare and contrast your competitors' capabilities with your own to ascertain any advantages or disadvantages in your current portfolio of systems.

### Observing Beyond Your Borders

- Whether through travel or the internet, study how similar companies in other countries design, develop, and deliver their products and services to uncover new practices that your industry and market can adopt.
- Attend trade shows and conferences that attract an international audience to observe how other regions conduct business.
- Watch product demonstrations and customer testimonials on online video platforms to identify innovative methods from markets outside your country.
- Follow industry peers and thought leaders from other parts of the world to discover new best practices.

# JIM RINALDI

Discovery Skill: **Networking**
Firm: **NASA Jet Propulsion Laboratory (JPL)**

Idea networking can take many forms. It can involve reaching out to individuals from diverse industries and backgrounds to identify potential solutions to your challenges. Networking for innovators may include seeking input in unconventional forums, including attending conferences and symposiums unrelated to your daily work. This requires curiosity about other professions and perspectives that may differ significantly from those found in your workplace and among your colleagues.

Jim Rinaldi is an IT innovator who demonstrated this curiosity as the CIO of the National Aeronautics and Space Administration's (NASA) Jet Propulsion Laboratory (JPL) from 2005 to 2019. During this time, the IT team received multiple recognitions, including the CIO 100 award. JPL conducts federally funded research and development for NASA, builds and operates robotic spacecraft and missions orbiting the Earth, and manages the NASA Deep Space Network. Rinaldi notes that JPL's motto, "Dare mighty things," is founded on experimenting with various approaches and viewing innovation from a different perspective. Rinaldi and his team embraced that motto and sought new ideas for IT innovation in unconventional places by practicing idea networking.

The last place you might expect the JPL IT team to look for innovative technologies for missions to the outer limits of our universe is the Consumer Electronics Show (CES), but that is precisely where they would go to conduct technology trend spotting. CES is an annual trade show in Las Vegas, Nevada, showcasing the latest product innovations from consumer technology firms.

Rinaldi and key staff members attended CES every year, where they walked the floors of the convention centers and focused on emerging technologies. He observed, "The first year that consumable 3D printing was on display, there was just a table where they had printed shoes, toys, and little trinkets. The next year, it was a room; after that, it was a hall." At that point, Rinaldi and his team recognized that this new technology could enable JPL engineers to innovate like never before. "We realized that you could 3D print an arm of a robot or a component from AutoCAD. Our engineers could effectively try things out with low-cost consumable 3D printing."

After recognizing the potential of this new technology, Rinaldi remarked that "IT gave these printers to engineering because they could do more with the technology than IT could. In return, we learned what they wanted and how they operated. We learned that people would line up for these technologies and would find new, innovative ways to employ them." To gain additional exposure to 3D printing and other emerging technologies, Rinaldi and his team created an Innovation Experience Center where people could come from all over JPL and NASA to try things out. They could see and experiment with some truly advanced technologies. People would come, learn, listen, and envision.

One outcome of deploying 3D printers at JPL was the ability to print components to assess the viability of an engineer's design long before an actual prototype was constructed. An engineer could print the component, review and modify the design, and then reprint it until it was ready to be built as a prototype.

Another use of the 3D printers Rinaldi and his team first discovered at CES was when the Mars Curiosity Rover encountered an interesting rock formation called Block Island. The rover took several pictures of the formation from different angles and then transmitted them back to Earth. A JPL team member developed a method to create a 3D-printed replica of Block Island that was so realistic it was eventually displayed at the Smithsonian Institution in Washington, D.C.

Networking for ideas, including emerging technologies such as 3D printing, has allowed Rinaldi and the IT team to gain recognition inside and outside JPL for their impact on the firm through IT-led innovation.

### Practicing the Skill of Networking

Networking for ideas can occur both physically and virtually. Here are a few additional examples of networking that demonstrate a wide range of options for practicing this skill.

- Contact local universities conducting research in relevant fields, such as materials science for manufacturing challenges or robotics for automation opportunities.
- Find out how your technology suppliers have assisted their customers in addressing challenges similar to the ones you are facing, regardless of whether they belong to the same industry.
- Engage with online communities, webinars, or virtual conferences when travel budgets are limited.
- Supplement in-person networking by searching for relevant podcasts or YouTube videos.
- Identify the skills your friends, neighbors, and professional colleagues possess that could benefit your organization, and invite them to share their knowledge. The following illustrates how one CIO accomplished this.

Steve Zerby is the former Senior Vice President and CIO of Owens Corning, a manufacturer of building materials including roofing, insulation, and composite products headquartered in Toledo, Ohio. During his tenure as CIO from 2013 to 2023, Owens Corning IT was recognized for its innovative accomplishments, earning the CIO 100 award seven times.

Regarding networking for ideas, Steve noted that "technology leaders in companies like ours need to have substantial personal networks that they can use as conduits to ideas, thoughts, and experiences." He cites the following example of how he utilized his network to educate his team and provide them with insight into new business and technology topics.

"I meet every month for a half-day with my top 30 to 40 IT leaders. One of our commitments was to use part of the meeting for education. I could come in and pretend that I know something, and the team will find that credible or not. Another option is to hire a consultant to create a workshop. But instead, I have utilized my network to help educate my leadership team. For example, I have had a professional colleague spend an hour with the team discussing his perspective on the future of digital payments. This is his expertise, so you can imagine there's immediate credibility because he knows what he's talking about. I trust him because I know him, so I know he won't come in with a wild set of ideas that aren't implementable. Our folks will go away with more education, and I suspect every one of them, for the next 30 days at some point, will look at some business problem through a different lens based on something they learned in that session."

By engaging a colleague to share his insights on digital payments, Zerby was able to educate his team about a technology that many had not encountered before. This session with a subject matter expert encouraged them to consider

incorporating it into their daily activities at Owens Corning in new
and innovative ways.

Jim Rinaldi and Steve Zerby have established structured methods for their teams to generate ideas through networking. Attending conferences outside their industry or connecting with a network of knowledgeable colleagues enables their teams to innovate for the companies they work for in an ongoing, sustainable manner.

## CHRIS CARUSO

Discovery Skill: **Associating**

Firm: **PPG**

Unlike the other discovery skills, associating is a cognitive process that involves linking two seemingly unrelated things to create an innovation. To illustrate this, let's consider how associating was used to find a new solution for an age-old customer problem.

PPG is one of the world's leading paints, coatings, and specialty materials manufacturers, and has been honored with the CIO 100 award five times. While the company produces coatings for use on a wide range of products, from appliances to airplanes and bakeware to bridges, one of the most recognizable applications of its products is painting the interior and exterior of residential and commercial buildings.

In the residential paint business, consumers begin the buying process by selecting the color they want to use for their painting project. For over a hundred years, the tools to assist consumers in choosing the right paint color have remained unchanged. Most paint companies use two primary methods to help customers pick a color: paint chips and liquid paint samples. As e-commerce gained popularity as a sales channel, many paint companies explored ways to sell their products online by focusing on digitizing these two-color selection options.

When the team at PPG considered how to tackle the challenge of online paint sales, they first reviewed their website and those of other paint manufacturers. They noticed that all featured similar content, including sections dedicated to color selection, products, and "where to buy" information. Website analytics revealed that digital color pages generated more traffic than any other section of the site.

While consumers prioritized viewing color on the website, there were two key challenges to turning that into online paint sales. First, in the online world, paint manufacturers face some of the same challenges typically encountered in physical stores. With many brands offering more than 1,000 color choices, the overwhelming variety can be daunting for online and offline customers. Another issue with selecting a color digitally is that each screen can display digital colors differently. As a result, this can lead to inaccurate "digital" color selection and, ultimately, a dissatisfied customer.

The IT team members began by networking for ideas to discover a new method to streamline the color selection process and address the screen color accuracy challenge. They searched TED Talk videos for similar challenges addressed in other industries or through new technologies. While watching a talk by Gary Flake, the founder and director of Microsoft Live Labs, they found exactly what they were hoping to discover.

In this video, Flake demonstrated a new way to browse online information that was under development, called Pivot. Pivot technology enabled users to visually browse and explore large amounts of data through image-based information. The demo at the TED Conference had an image of every available *Sports Illustrated* magazine cover loaded into the Pivot tool, along with the metadata associated with each edition. Flake's screen was filled with hundreds of tiny images of magazine covers that were discernible when he began to apply filters that would remove covers that didn't meet the selection criteria. The filters included the sport, the name of a featured athlete, a team, a sporting event, or a specific publication date.

As the user specifies the filters they want to apply, the covers that don't meet the criteria disappear, and the remaining images enlarge and appear to move forward on the screen. The same occurs when the user zooms in on a particular magazine cover; a window appears with all the metadata about that magazine and hyperlinks to other editions containing the same metadata.

When the team from PPG viewed the Pivot demo, they immediately made the association between the on-screen display of hundreds of magazine covers and the hundreds of color chips in racks at paint stores and home centers. They connected filtering magazines by sport, athlete, etc., with sorting through paint colors by color family (reds, greens, blues), color groups (clean, pure, shaded, and muted), and color popularity. The popularity of a specific color was one of several new pieces of metadata the team created based on the actual volume of paint sold for that color.

The initial version of the PPG Pivot tool was completed just as their largest customer, a home center based in the United States, requested their help selling liquid paint samples online on their website. When the team demonstrated the tool to this customer, they were so impressed that they immediately asked that it be implemented as quickly as possible. In just a few weeks, the Pivot tool was integrated into the home center's website and was heavily promoted by the retailer.

The team filed for and received a U.S. patent on their unique color selection process. Additionally, the color selection tool was implemented for several PPG paint brands in the residential and automotive refinish coatings sector to assist collision centers in finding the paint formula that matches a vehicle's color.

By connecting two seemingly unrelated activities, searching for a magazine cover and choosing a color for a painting project, the team at PPG demonstrated that associating can lead to new and improved ways for their customers to purchase paint.

## Practicing the Skill of Associating

This success at PPG illustrates a key point: associating stands apart from other discovery skills. Developing this skill requires establishing a mental habit of combining diverse inputs to create something entirely new and original. These inputs often result from questioning, experimentation, observation, and networking. Consequently, practicing all five discovery skills together increases your chances of discovering your desired innovation.

To practice associating, start by taking two seemingly unrelated technologies and imagining what would happen if you could leverage the best of each to create something new. Imagine you own a window replacement business and want to find a way to increase sales in a specific neighborhood. You might consider what would happen if you combined two different online services: mapping that provides 360-degree views from the street level (e.g., Google Street View or Microsoft Bing Streetside) and a large language model (LLM) chatbot (e.g., ChatGPT or Claude) to identify new customers. What if you could upload a Street View photo to ChatGPT with the prompt: "Do the windows in this house need replacement?" Suppose the LLM could analyze the image and successfully determine whether that house's windows need replacement. Could you combine both capabilities to analyze neighborhood homes systematically using the LLM and generate qualified customer leads? Practicing associations like this could lead to innovative and novel ways to generate sales and grow market share.

Whether assisting homeowners in choosing a paint color or guiding a window replacement company in acquiring new customers, you must challenge yourself to think outside the box. Breakthrough ideas are often formed by finding innovative solutions by connecting things that have never been connected before.

## CHAPTER 3

# Integrating All Five Discovery Skills

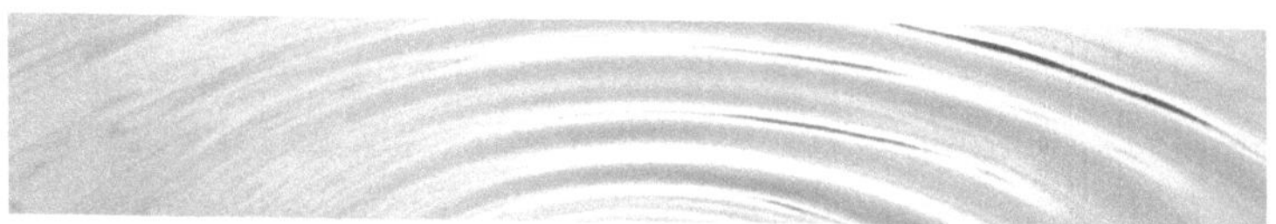

The five examples of discovery skills presented thus far are individual steps in the innovation journey. However, the survey showed that CIO 100 award-winning organizations foster more of these skills than their peers. The combination of these five skills enables breakthrough innovation by creating a systematic approach to innovation.

Questioning uncovers problems and challenges assumptions, but without observing, you miss the unique distinctions of the circumstances that make solutions relevant. Observing uncovers patterns and unmet needs, but without networking, you lack diverse perspectives to identify the most appropriate solution. Networking exposes you to new ideas and approaches, but without experimenting, those ideas remain theoretical. Experimenting generates data and insights, but without associating, you can't connect the dots to create a novel solution. Each skill fills the gaps left by the others.

The following story from PPG offers a broader perspective on the purpose, scope, and vision for technology innovation and the journey, including its challenges and successes. Throughout this story, examples of how the team at PPG practiced the five discovery skills will be illustrated.

## PPG's Digital Paint Purchase Process: An Innovation Story

Architectural paint has been used for thousands of years, with its origins dating back to the early Egyptians and Greeks, who employed it to decorate walls and other surfaces. In the 19th century, manufacturing and selling this paint became more common with the rise of industrial production methods. Over the past century, advancements in paint technology and color science have transformed the industry into a multi-billion-dollar global market.

Despite advancements in the product inside the can, purchasing paint has remained unchanged for over one hundred years. The first step in the buying process begins with selecting a color for your project. Dating back to 1875, manufacturers produced pamphlets for consumers and painters, featuring color swatches of paint. After selecting their preferred color, they could buy a small amount of paint to apply to their wall to confirm their decision. More than one hundred years later, that process remains fundamentally unchanged. While the quality and quantity of color sampling materials have improved, so has the complexity of decision-making due to the wide range of color options and the variety of special-purpose paints.

## Establishing A Strategy to Change
## How Paint Is Purchased

In early 2009, a small group of IT leaders in PPG's North American Architectural Coatings business set out to challenge the conventional wisdom regarding how consumers made their paint purchasing decisions. The team included the global IT director for all of PPG's architectural coatings businesses, the IT director for the North American business unit, and the digital marketing IT manager responsible for North American customer-facing digital initiatives, such as mobile apps, websites, and social media. Their vision, called the "digital paint purchase process," outlined the steps customers took when buying paint; however, the key distinction was that it was a fully digital experience.

Those steps consisted of selecting colors, building color confidence by viewing the colors on your desired surface, choosing the right paint, determining the quantity and sheen required for the project, and finally, purchasing the product. This group of IT leaders envisioned a seamless virtual process enabling consumers to create and access their painting project information from home using personal computers, smartphones, or a retail store at a kiosk. Finally, they envisioned new ways to assist consumers in their paint selection process, including easy-to-use color visualization tools, paint color popularity trends, and integration with social media to gather feedback from friends and family.

Though they lacked the budget to develop their vision, they created a roadmap, including a one-page visual representation of their vision, which would serve as the foundation for their strategic and tactical plans over the next several years. This document would communicate their strategy and obtain management support for the funding needed in the following year's budget.

In May of that year, a significant event occurred that would dramatically alter the trajectory and velocity of the team's vision. A May 20, 2009, *New York Times* article titled "On Your iPhone: Point and Paint" announced the launch of an app from one of PPG's North American competitors, Benjamin Moore. The Color-Capture Ben app "allows users to match any color in a photograph to one of Benjamin Moore's more than 3,300 paint colors." After downloading the app and taking a picture, the user could touch their iPhone screen to the photo area to highlight a specific color, and then the app would find the closest color in the Benjamin Moore palette. They could then save the color and find the nearest distributor of Benjamin Moore paint using the iPhone's GPS.

Shortly after the release of this app, Sherwin-Williams, another major competitor, launched a similar app for the iPhone. The exposure these two competitors gained prompted PPG's leadership to ask the IT team when they would have a comparable app. The team understood that developing an iPhone app with the same functionality would not help them achieve their vision to "change how consumers purchase paint" and could hinder their progress in creating the differentiated solution outlined in their roadmap. They also knew that the cost of building the app exceeded $100,000, surpassing any discretionary funding they had.

## Creating the Opportunity for Change

Over the next several months, the IT organization continued to receive requests to respond to the competition. In the fourth quarter of 2009, the team determined they needed to develop an affordable alternative solution to align with their overall strategy and roadmap. Given that the iPhone had been on the market for only a few years and accounted for a small fraction of users relative to personal computer owners, the team felt they could take a different approach. They opted to build a desktop app that could reach more consumers, facilitate the seamless buying process they envisioned, and be developed at a fraction of the cost of a mobile app.

## Experimenting by Prototyping

The team utilized the discovery skill of *experimenting* by creating a prototype of the app using PowerPoint slides to illustrate its functionality. The app would run on Windows and Apple computers, allowing users to click on any image on their screens to find the color that most closely matches a shade in the PPG palette. This would enable a user of the app to visit any website, such as an online home décor magazine or retailer, and find color inspiration, including furniture, rugs, or wallpaper that coordinates with a PPG color. They could then save their favorite colors in the app and find the nearest paint location to purchase their paint.

The app's most critical and differentiated feature was linking to an online visualizer, allowing users to build confidence in the selected color. With just a few clicks of their mouse, a consumer could view a color on any website and instantly see how it would look in a room setting. The team knew that building this would enable the consumer to complete the first two steps in the "digital paint purchase process."

The next step was to review the PowerPoint slide prototype and the one-page roadmap with the business's vice president and staff managers. During the discussion, it was noted that a desktop app could potentially reach a larger audience than an iPhone app and would cost significantly less. However, finding funding for the project outside the IT budget would require approval.

The staff managers said they liked the idea but preferred an iPhone app. They informed the team that they would prefer not to fund a desktop app. However, the VP of the business was inspired by the originality of the idea and its potential to set it apart from other paint companies. He recognized that while it was the fourth quarter of the year and there wasn't much left in the IT budget, he would support the team if they could find the funding to develop the software. The North American IT Director projected that around $30,000 would be available to spend from his budget, which was considerably lower than their initial estimations for the product's development. Therefore, they opted to create a beta version of the app, utilizing the available budget.

## Finding Support to Fund the Vision

A few weeks later, in January 2010, the IT Manager and his digital marketing team worked with a small group of external resources to develop the app's beta version. It was then ready for demonstration to the VP and staff managers. After the demo, the feedback was positive, but the staff managers still wanted an iPhone app. The VP interjected that they planned to meet with their largest customer in the next few weeks, which would be an excellent opportunity to get their opinion. The team customized the app with the retailer's color palette and named it "ColorClix™"

During the customer meeting, ColorClix™ was on the agenda during lunch, when the account team thought they could demo the app and generate feedback. The response was far better than the account team had anticipated. The customer loved the ColorClix™ app and wanted it available for download as quickly as possible. To meet the expectations of this important account, the digital marketing IT team developed the full version and launched it in February 2010. Six weeks later, they initiated a promotional campaign for the tool.

In April, *USA Today* featured ColorClix™ on the front page of both the print edition and its website homepage. The print edition alone had a circulation exceeding 2 million readers. This was followed by a full-page advertisement in *Better Homes & Gardens*, which had a circulation of 5 million. Lastly, and most impressively, ColorClix™ was featured in a promotional flyer distributed with 50 million Sunday newspapers nationwide.

## Moving Forward with the Roadmap

With momentum for the "digital paint purchase process" growing, the IT leaders began to move forward with the launch of the next phase of the roadmap. While the development of version 2 of the desktop app, with new features including capturing multiple colors from an image, sharing project information via social media, and viewing designer-selected colors, was underway, the team recognized the need to tackle the next step in the buying process: building color confidence.

Visualizing color in a room setting is a crucial digital capability that helps consumers feel confident in their color selection. PPG already had a standalone online visualizer that enabled the digital "painting" of stock photos of various room settings, including living rooms, kitchens, and bedrooms. The technology team's challenge was integrating the colors the desktop tool selected with the web-based visualizer. From start to finish, the team only took a few months to integrate the two applications and release the functionality to their customers.

The team's next step was to enable color selection "on the go." This meant developing mobile apps for iPhone and Android smartphones that could synchronize paint project information with the desktop and web applications. PPG introduced its smartphone applications nearly two years after its competitors launched their mobile apps. The key difference was that the PPG app distinguished itself from all other competitors due to its integration with the entire suite of color tools developed by the IT team.

This differentiation was highlighted in a *Wall Street Journal* article on April 14, 2011, titled "Narrowing 3,500 Paint Choices to 1 Perfect Hue." The article highlighted smartphone apps from several paint manufacturers, including PPG, Sherwin-Williams, and Benjamin Moore. The article's author discussed the story in an online *WSJ* video titled "Digits: Apps for Picking Paint Colors," which outlined the author's favorite apps. She specifically called out the ColorClix™ mobile app for integrating the mobile app and the website and sharing color information between the two.

As PPG's strategy received increased media recognition for helping its customers select colors for their painting projects, the IT organization was also recognized with the CIO 100 award later that year for its work delivering this suite of integrated applications. This helped validate and reinforce the strategy, providing internal support to expand the capabilities of these tools and accelerate their delivery.

## Building Digital Color Confidence

With the core color selection apps launched, it was time to consider developing a more robust color confidence tool. Specifically, PPG's consumer research showed that their customers wanted to visualize the color in their rooms rather than in a stock photo. This capability was significantly more complex than their current room visualizer and would require a substantial investment to deliver a solution that met their customers' needs.

## Observing and Questioning to Solve the Visualization Challenge

The team reviewed their competitors' paint visualizers and observed how users interacted with the software. These tools were developed to mimic the real-world process of applying painter's tape to distinguish areas of the room you don't want to paint from those you do. The user could only use a drawing tool that creates straight lines, similar to painter's tape, to differentiate a wall surface from furniture, the floor, or windows.

The team noted that most rooms rarely contain items that can be separated in a photo by drawing a straight line. Home décor items, such as furniture, mirrors, and lampshades, often feature rounded corners, making them difficult to distinguish from a wall when using a virtual representation of painter's tape. After a considerable effort to carefully draw lines on the image of their room, they could finally digitally paint with their chosen colors.

The IT leaders wanted to challenge the status quo in online paint visualization by *questioning* the state of the art. They asked, "Why hasn't a visualizer already been created to automatically detect different objects and surfaces, like walls and ceilings, in a room image and separate them to allow for painting the surfaces while leaving other items untouched?" Was it a limitation of the technology, or was it simply an assumption that users would tolerate the inconvenience? They envisioned a tool that would be less frustrating and more intuitive for their customers to use. However, they knew they would need assistance creating this highly complex and unique visualization technology.

## Identifying Solutions by Networking for Ideas and Associating

PPG has established relationships with leading universities worldwide to collaborate on research, education, and community engagement initiatives. One such university, Carnegie Mellon University (CMU), is located in the same city as PPG's headquarters in Pittsburgh and is renowned for its computer science and robotics programs. The North American IT Director decided to reach out to the company's contacts at CMU to help him network and gather ideas for building this first-of-its-kind visualization technology.

During his visit to the campus, he met a robotics department professor who was researching image recognition technology. This research focused on intelligent transportation, underwater and medical imaging, and computer vision to estimate grape production in vineyards. His work to help wine producers determine the size of that year's harvest led PPG to ask him for assistance in developing a next-generation paint visualizer.

This researcher had developed an algorithm that would process an image of a grapevine and count the number of grapes in the picture. To accurately do this, he had to develop the ability to separate a green grape from the green leaves on the vine. This capability prompted the IT Director to connect the grape counting solution with the virtual room painting challenge. By linking the separation of green grapes and leaves in one application to the separation of home décor and surfaces in another, he demonstrated his ability to utilize the *associating* discovery skill.

## The Innovative Breakthrough

While this researcher lacked experience in paint color visualization, his expertise in developing algorithms for applications such as grape counting, medicine, and underwater environments was precisely what the team needed. With his expertise, they created a visualizer that could automatically separate an image into two categories: surfaces to be painted and surfaces not to be painted, thereby simplifying the user experience.

To bring their innovation to market, the team integrated the new room painter technology with the desktop application and launched it as part of an online project center. This digital platform enabled consumers to manage multiple painting projects and track associated paint colors. Recognizing the uniqueness of their approach to color visualization, PPG filed and received a patent for the technology.

The online project center, featuring the visualizer, quickly gained industry recognition. The platform was nominated for a Webby Award, often called "The Oscars of the Internet," in the "Lifestyle Website and Mobile Sites" category. Although the award ultimately went to *Martha Stewart Living*, the nomination reflected PPG's increasing leadership in innovation within the digital home improvement space.

## Completing the Digital Challenge

The final step in completing the roadmap was to develop a digital capability for in-store use that integrates with the existing suite of color tools. This phase would be the most complex and capital-intensive part.

The goal was to develop an in-store kiosk integrated with a new color chip rack under development for paint distributors. This solution features a 42-inch touchscreen, a barcode reader for scanning paint color chips, and internet connectivity, enabling the collection of point-of-sale data and transmitting updates to the kiosk. The kiosk design allowed a retail customer to select a paper paint chip from the rack, scan the barcode, and interact with the color on-screen in various room settings. If the customer wanted to see the color at home, they could send an email from the kiosk with a link that automatically transmitted the color to the online room painter, allowing them to visualize how it would look in their room.

Upon completion of the development, the kiosks were constructed, loaded with the software, and installed in more than 700 retail locations across the country. As a result of this effort, PPG was recognized again with the CIO 100 award for developing this industry-first retail technology. Additionally, two patents were filed and granted for future capabilities that utilize the kiosk to help consumers find coordinating home décor products, including paint, and to analyze user behavior with the kiosk to aid in cross-selling other products in the store.

## The Vision Realized

Five years after launching the initiative, the IT team achieved a significant milestone with the rollout of in-store kiosks, completing the core elements of their innovation roadmap. Along the way, PPG secured four patents and earned multiple industry awards and accolades. Their success was driven by consistently applying the five discovery skills: experimenting, questioning, observing, networking, and associating, combined with a steadfast commitment to a shared vision. By focusing on their strategy and staying true to their goal of "changing the way that consumers purchase paint," they proved that even resource-constrained organizations can deliver breakthrough innovations that can transform an industry.

## CHAPTER 4

# Enabling Innovation by Listening, Communicating, & Acting Courageously

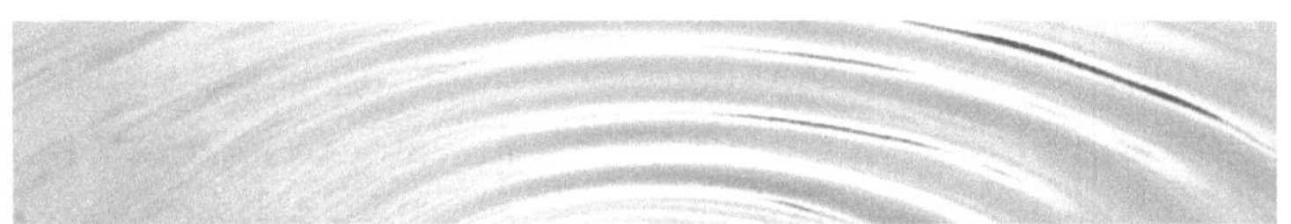

> "When there was a challenge or an extreme
> event, that was when I ran into the fire."
> **—Renee Zaugg**

In addition to fostering the development of innovation skills within their organization, impactful leaders exhibit three key qualities: *listening to learn, communicating to connect,* and *acting courageously to do the next right thing*. The following story from Renee Zaugg, the former CIO of Otis Elevator, illustrates how all three of these qualities were essential in helping her navigate the challenges of starting with a new firm, separating it from its parent company, and leading her team through a global pandemic. After a crash course in learning about the elevator business, building trust with her staff, and navigating the unknown of COVID,

Zaugg and her team were rewarded with the CIO 100 award in 2022 for their innovative Otis One™ application.

## How Otis Elevator's CIO Built Trust and Drove Innovation through Crisis

Zaugg's 38-year healthcare career began as a third-shift tape hanger and ended as Senior Vice President of Integrated Enterprise Infrastructure. She built her reputation by consistently "running into the fire" to solve critical issues, becoming known for achieving results through persistence, operational knowledge, and business acumen.

In January 2020, Zaugg had just started to settle into retirement when she was approached regarding the position of Chief Information Officer at Otis Elevator Company. After more than 160 years of operations, the global elevator manufacturer was separating from its parent company, United Technologies. The newly independent company needed an IT leader with experience in infrastructure, applications, and systems architecture, skills it had previously depended on United Technologies to provide. Zaugg had the skills they sought and accepted the job, unaware that COVID would soon force her to "run into the fire" again.

## Courageous Leadership: Seizing Opportunity in Crisis

Courageous leaders don't wait for perfect conditions; they identify what is right for the organization and act decisively, even when it necessitates difficult conversations about investment and change.

With the upcoming separation from United Technologies and the start of the COVID pandemic, Zaugg recognized the urgency to proceed quickly, prioritizing the evaluation of the company's installed technology. It was immediately clear to her that it required updating to meet the business's needs, and the separation of Otis from the parent company would create the opportunity for the essential financial investment to make these improvements. Although it may have been easier to avoid asking for the investment needed to upgrade the company's infrastructure, Zaugg understood that it was both the right thing to do and the right time to do it. She decided to seize the opportunity and think like a startup that had been around for more than 160 years. She knew her team would embrace this approach and would be eager to implement new technologies and updated methods for managing them.

One of the first investments required was implementing videoconferencing technology for employees in China. Before the pandemic, Otis had video conferencing licenses for 400 employees. However, with the shift to remote work necessitated by the pandemic, they quickly increased the total to 8,000. Acquiring and implementing the software required the IT team to work around the clock. Although the work was demanding, they felt a sense of purpose in the role they played in supporting the business during such a crucial moment. They knew that Zaugg would brag about the team to their business colleagues, which, in turn, would elevate the IT organization's stature within the company.

### Key Insight and Consideration:

Zaugg's decision to push for infrastructure investment during a crisis illustrates a key principle of courageous leadership: the "next right thing" often requires short-term difficulty for long-term benefit. Many leaders avoid making tough requests during uncertain times. Still, brave leaders recognize that crises frequently create unique windows of opportunity for necessary changes that would be harder to justify in normal circumstances. The key is distinguishing between what's urgent and important, then having the conviction to advocate for the important decisions even when they're inconvenient. Zaugg's team responded positively not just because of the challenge, but because they could see their leader was willing to take personal risks to position them for success.

*What "next right thing" have you been avoiding because it seems too difficult to request or implement? How might current challenges create an opportunity for necessary changes?*

## Listening to Learn and Build Trust

One of Zaugg's early priorities was to build strong relationships with her team by getting to know them and allowing them to know her. At her first staff meeting, she shared her "entourage tapestry," a collage of photos representing the people who inspired her. Among them were her parents, who raised 14 children, including three sets of twins, a former mentor, several family members, and Mother Teresa. By sharing this deeply personal artifact, Zaugg communicated her core values and demonstrated that vulnerability was a cornerstone of her leadership approach. It gave her team a meaningful glimpse into who she was beyond the title, helping to build trust and connection from the outset.

Zaugg then asked each team member to create and share their tapestry with her. The request was initially met with skepticism: "The house is on fire, Renee, and you want to spend time at staff meetings doing this?" someone asked. Her response was clear and firm: "Absolutely. We can't do this unless we are all good leaders." Despite the urgency of their work, the team embraced the exercise. What followed was a powerful, unifying experience. As each person shared the stories behind their tapestry, team members better understood one another's backgrounds, values, and influences. It became an opportunity for reflection, connection, and gratitude, creating an environment where people could honor those who had shaped their lives. This approach provided Zaugg a valuable chance to *listen and learn* about her team and what mattered most to them. It also enabled her to build trust among and between her team members, preparing them for the challenges ahead.

> **Key Insight and Consideration:**
> This approach demonstrates how vulnerability-first leadership fosters organizational trust. When leaders share personal stories first, they allow others to feel safe being authentic, enhancing the team's performance. The key is to make vulnerability purposeful. Zaugg didn't just share personal details; she shared what shaped her values.
>
> *What structured approach could you use to learn what motivates your team members? How could understanding their core influences change how you lead them?*

## Communicating to Connect with the Organization's Purpose

Now that she had established an initial understanding of the organization, its people, and its challenges, Zaugg and her team were positioned to focus on driving meaningful business impact.

One initiative already underway before Zaugg's arrival was Otis One™, a project designed to provide Otis and its customers with valuable insights into elevator performance and health. The goal is to enable predictive maintenance by identifying service needs before an elevator is taken out of service. To achieve this, real-time elevator status must be accessible to key users, including property owners and field engineers.

For a project like Otis One™, the team typically visits the Otis test towers to gain firsthand insight into how elevators operate, which is critical for designing an effective solution. However, restrictions caused by the pandemic limited site access to essential personnel only. This presented a unique leadership challenge for Zaugg: helping the team grasp the strategic importance of Otis One™ while fostering a deeper connection to the company's mission. In her previous experience in healthcare, it was relatively easy to align employees with a clear purpose: preventing disease and saving lives. Inspiring the same level of purpose in the elevator industry required a more intentional approach.

Zaugg realized that Otis was doing far more than merely building and servicing elevators and escalators. With more than two billion people relying on their products daily, safety and reliability weren't just features; they were expectations. She began to view elevators as essential infrastructure, especially in urban environments where most buildings would be inaccessible above the fourth floor without them. Although she knew Otis was the original inventor of the elevator, she hadn't fully considered the profound impact if those elevators failed to function.

With this new perspective, Zaugg crafted a narrative to communicate to her team the significance of Otis One™ and connect more deeply with the project's mission. She asked them to imagine a family trip to Disneyland without elevators. How would parents with strollers get around? What about individuals in wheelchairs? How would they access the attractions? For Zaugg, the goal was to enhance the team's sense of purpose and cultivate a connection by framing their work as essential to mobility and safety. She collaborated with the communications team to spread this message, eventually embedding it throughout the company, from frontline functions to executive-level leadership.

### Key Insight and Consideration:

Zaugg's approach illustrates a fundamental principle of inspirational leadership: connecting complex technical work to human impact through powerful, relatable stories. Instead of detailing the technical benefits of predictive maintenance, she illustrated the consequences of elevator failure in a way that evoked genuine emotional understanding.

This technique, shifting from features to human stories, works across industries because it taps into our natural empathy and desire to make a difference. The most effective leaders don't just communicate what their teams do; they help people understand why it matters to real human beings.

*What human story could you tell about your team's work? How can you help them see the real people impacted by their actions each day?*

## Moving forward with Otis One™

With a clear understanding of customer needs, the team was ready to begin developing Otis One™. The goal was ambitious: create a system that could proactively detect elevator issues and notify a global network of over 65,000 technicians. Designing a solution that worked across multiple generations of elevator technology posed a significant challenge. It involved standardizing and transmitting large amounts of data from elevators worldwide, regardless of age or model, to enable integration with the company's ERP system.

The final product featured a customer-facing portal that displays elevator locations, facilitates service requests, and provides real-time monitoring capabilities. Internally, Otis employees can track performance data, notify customers of required maintenance, and utilize predictive analytics to recommend preventive actions or system upgrades, bringing the company's vision of intelligent elevator management to life.

## The Journey to Innovation through Listening, Communicating, and Courage

As noted in Otis' 2022 annual report, Otis One™ was identified as one of the company's top priorities and was operational in 30 countries and territories. Two years after she joined the company, Zaugg and the Otis One™ team received the CIO 100 Award for the system. This honor reflected the success of a transformative technology initiative and the resilience and leadership required to achieve it.

In the face of extraordinary challenges, including a global pandemic, separation from a long-time parent company, and the urgent need to modernize infrastructure and IT practices, Zaugg led with clarity, empathy, and conviction. She quickly immersed herself in the business, built trust within her team, and helped employees connect their work to a greater purpose. Perhaps most importantly, she had the courage to consistently do the next right thing, positioning Otis for long-term success and exemplifying purpose-driven leadership.

Zaugg's story illustrates an important point: leading technology-driven innovation involves more than technical skills or business acumen. Across industries and organizations of all sizes, the leaders who make the most significant impact are those who listen with intention, communicate to build connection, and act courageously when it matters most. These qualities characterize transformational leadership and are essential for creating authentic and enduring change.

# Impacting the Firm: What Did We Learn

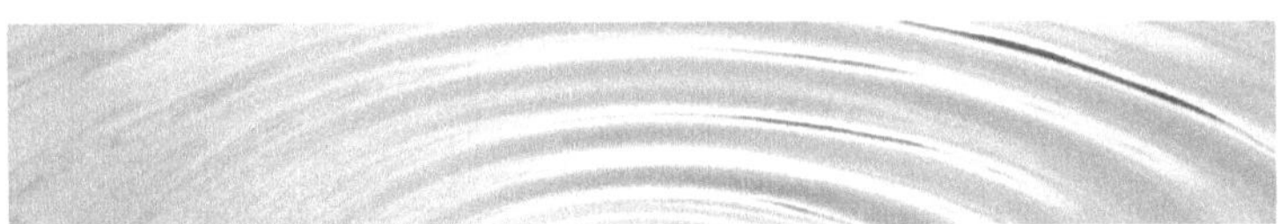

## Technology Driven Innovation: Impactful Leadership Lessons

Innovation is not a luxury for technology leaders; it is the defining characteristic that separates those who shape the future from those who merely react to it. As you've read throughout this section, the most impactful technology leaders don't wait for perfect conditions, unlimited budgets, or executive mandates to drive meaningful change. They understand that innovation skills can be learned if fostered within any organization, regardless of size, industry, or resource constraints.

## The Innovation Imperative

The stories shared in this section, from Lexmark's Focus to Future hackathons to PPG's digital paint color technology to NASA JPL's consumer electronics insights, demonstrate a fundamental truth: innovation happens when leaders intentionally create the conditions for it to thrive. These leaders didn't stumble upon breakthrough ideas by accident. They systematically practiced the discovery skills that distinguish innovative leaders from their peers.

The survey data points to the importance of discovery skills. Technology leaders from award-winning organizations strongly focus on all five skills, with the most significant emphasis on experimenting (+15%) and associating (+11%). More importantly, 50% of CIO 100 leaders foster all five skills in their organizations, compared to 37% of their counterparts.

## Beyond Technical Skills: Essential Qualities of Innovative Leaders

Technical skills alone don't create transformational change. As Renee Zaugg's journey at Otis Elevator illustrates, lasting innovation requires three essential leadership qualities: listening to learn, communicating to connect, and acting courageously to do the next right thing. These qualities enable leaders to build trust, shared purpose, and organizational courage to pursue breakthrough ideas despite uncertainty.

Innovation demands vulnerability, the willingness to experiment and potentially fail, to ask uncomfortable questions that challenge the status quo, and to connect seemingly unrelated ideas in ways that others may consider unconventional. It requires leaders who can see beyond immediate operational pressures to identify opportunities that others miss, and who possess the conviction to advocate for necessary investments even when they're difficult to justify.

## Your Innovation Journey Begins Now

The choice before you is simple but not easy: Will you continue running on the technology treadmill, constantly reacting to the next new thing while struggling to keep pace? Or will you step off that treadmill and become an innovation driver who shapes your organization's future?

This transformation begins with a single decision to view innovation not as an additional responsibility but as an opportunity to have a lasting impact on your firm. It means allowing time for questioning current assumptions, creating opportunities for experimentation, and building networks that extend far beyond your immediate industry and expertise.

**Consider these actionable steps as you begin:**

- **Start small, but start now.** You don't need a large budget or dedicated innovation lab. Begin by consistently implementing a single discovery skill. Schedule monthly strategy sessions to ask "why" questions about current business challenges. Attend a conference outside your industry. Create a simple prototype using existing tools. The key is consistent practice, not perfect conditions.

- **Make innovation visible.** Like the leaders profiled in this section, communicate your innovation efforts and celebrate successes and learning experiences. Help your team understand that innovation is everyone's responsibility, not just R&D's domain. Share stories that connect your technology work to human impact and organizational purpose.

- **Get personally involved.** Don't delegate innovation leadership, own it. Show your team that innovation matters by actively participating in idea sessions, giving thoughtful feedback on prototypes, and leading firsthand visits to your company's operations or customers' environments. Your involvement sets the tone and signals innovation as a shared, high-priority commitment.

## Innovation Leadership Demands Perseverance

The accomplishments featured in this section didn't happen overnight. PPG's digital paint purchase journey took five years to complete. Lexmark's Focus to Future evolved after multiple iterations before becoming a source of breakthrough innovations. NASA JPL's consumer electronics insights required years of consistent attendance at industry conferences before yielding transformational ideas.

This timeline should encourage, not discourage. It should help reinforce that developing a unique innovation requires time. Every time you experiment, ask a probing question, or make a revealing observation, you are making progress toward your overall innovation goals.

## Pivoting to Innovation Leadership

Operational excellence is a fundamental requirement of any technology leader; it isn't a differentiator. Maintaining and upgrading systems, delivering projects on time and under budget, and securing your firm's assets are expectations of the job.

What distinguishes you as a leader is your willingness to go beyond daily operations and address the hidden challenges that hinder progress for your company and your customers. In your role, you can reshape how your organization tackles challenges by helping your team see problems as opportunities for change and championing bold, breakthrough ideas.

Innovation distinguishes between a leader and a follower, as Steve Jobs observed. But more than that, innovation distinguishes between leaders who manage the present and those who create the future. The choice, and the opportunity, is yours.

## Impacting the Firm: A Final Reflection

Do you remember what you wanted to be when you were a child? When I was eight, I dreamed of becoming an inventor. I loved taking things apart and putting them back together, sometimes with disastrous results, and other times creating something entirely new. Once, I took apart an old radio and turned it into a loudspeaker, much to my parents' dismay. When I told them I wanted to play football, they agreed, on one condition: I could only play if I became a placekicker, to minimize the risk of getting hurt. Rather than waiting for them to buy me a kicking tee, I built my own from scraps of wood I found in the basement of our home. Now, when our grandchildren visit, I see that same spark of imagination in them as they turn blocks and boxes into castles and rocket ships; dreaming, building, and discovering the joy of creating something from nothing.

I witnessed this same sense of wonder and creative power in the university students I advised at Carnegie Mellon University. These third and fourth-year students worked with nonprofit organizations on projects ranging from meal services to youth mentorship programs. Within only four months, while juggling multiple classes, they became passionate ambassadors who designed and delivered fully operational systems that made a real impact for their partner organizations.

But here's the troubling question: where does this innovative energy go once new graduates join our companies? Are we unknowingly crushing the creativity and fresh thinking we want through rigid technology standards, extensive policies, and entrenched ways of doing things? How frequently do we silence promising ideas from newcomers with phrases like "that would never work here" or "we tried that before"? Just as those university students transformed into passionate advocates, how can you create an environment where your team members approach challenges with the same curiosity and commitment to make a meaningful impact on your firm?

# PART TWO

# IMPACTING THE TEAM: CREATING A GREAT WORKPLACE

## CHAPTER 6

# Impacting the Team: Why It Matters

"You manage things; you lead people."
**—Grace Murray Hopper**

This simple but profound reminder from computing pioneer Grace Hopper underscores a truth often overlooked in technology leadership: your role is not to lead technology but to lead *people*. Many technology professionals rise through the ranks based on their technical expertise, only to find that leadership requires a different skill set. This transition can be unsettling, but it shouldn't be discouraging. Leadership is a journey of growth. Those who acknowledge the gap between technical skill and people leadership, and who actively seek to bridge it, are the ones who ultimately thrive.

Great technology leaders understand that their most significant responsibility is not managing systems, but developing people. They create environments where individuals are empowered to grow, contribute, and thrive. These leaders take time to understand what motivates their team members, professionally and personally. They ask questions like: *Does this individual aspire to lead others, or are they most fulfilled as a deep subject matter expert? What experiences or training would help this person reach their goals? Are they developing too quickly without the proper support, or not fast enough to stay engaged?*

Effective leaders are observant, responsive, and adaptive. They tailor their guidance to each person's pace and potential. Most importantly, they understand that leadership involves supporting others in achieving success and motivating their team members to reach their full potential.

## Herzberg's Two-Factor Theory of Motivation

A significant amount of research has been conducted on how to motivate employees. One well-known motivation theory is Herzberg's Two-Factor Theory, which was developed by Frederick Herzberg in 1959. Herzberg interviewed employees to understand what they liked and disliked about their jobs and identified two key aspects of workplace job satisfaction. He refers to these two dimensions as hygiene factors and motivators.

Hygiene factors are external to the job and can prevent employees from feeling dissatisfied. These include salaries and benefits, the quality of their supervisor, company policies, relationships with coworkers, and working conditions. Motivators, inherent to the job itself, can foster employee growth. These include the quality of work assignments, employee development, expanding employee responsibilities, acknowledgment of accomplishments, and career advancement to higher levels within the organization.

## Motivation Practices of Technology Leaders

Drawing on Herzberg's theory of motivators and hygiene factors, a survey was conducted with responses from more than 800 technology leaders across 53 countries to explore how they motivate their teams. Each participant was asked, "What are the five most important areas you focus on to create a great workplace?" The ten options comprised five hygiene factors and five motivators; however, survey participants were not informed that the question was based on Herzberg's theory. The responses were analyzed across multiple dimensions, including country, gender, industry, and leadership level, and are detailed in Appendix B.

**Figure 4-1** shows the survey results ranked and grouped into high, medium, and low-importance categories. In the high-importance group, 67% or more of the leaders chose three motivators (recognition, meaningful work, and employee development) and one hygiene factor (salaries and benefits).

| What are the five most important areas you focus on to create a great workplace? | | |
|---|---|---|
| **Discovery Skill** | **% of All Respondents** | |
| Recognition | 72% | } High |
| Assigning meaningful work to employees | 70% | |
| Employee development | 68% | |
| Salaries and benefits | 67% | |
| Positive coworker relationships | 52% | } Medium |
| Quality of supervisors | 52% | |
| Employee career advancement | 49% | |
| Working conditions | 34% | } Low |
| Expanding employee job responsibilities | 23% | |
| Company policies and procedures | 8% | |

| Motivator | Hygiene Factor |
|---|---|

*Figure 4-1*

The medium-importance group includes two hygiene factors (positive coworker relationships and quality of supervisors) and one motivator (career advancement), with all responses clustered around 50%, ranging from 49% to 52%. Finally, the low-importance group consists of the hygiene factors of working conditions and company policies, along with the expanded job responsibilities motivator, each selected by 34% or fewer respondents.

Although these overall patterns are insightful, examining the data based on leadership levels reveals even more detailed insights. **Figure 4-2** shows the aggregated and ranked results for senior executives with titles such as CxO (e.g., Chief Information Officer, Chief Technology Officer, Chief Digital Officer), Vice President of Information Technology, or Head of Information Technology, along with respondents in CISO and Director/Manager roles.

| What are the five most important areas you focus on to create a great workplace? | | | |
|---|---|---|---|
| Factor | Senior Executives | CISO | Director / Manager |
| Salaries and benefits | 65% | 75% | 82% |
| Quality of supervisors | 52% | 49% | 50% |
| Company policies and procedures | 8% | 4% | 7% |
| Positive coworker relationships | 52% | 54% | 61% |
| Working conditions | 34% | 37% | 36% |
| Employee development | 68% | 72% | 61% |
| Assigning meaningful work to employees | 71% | 68% | 75% |
| Expanding employee job responsibilities | 23% | 27% | 18% |
| Recognition | 75% | 65% | 64% |
| Employee career advancement | 49% | 46% | 50% |

| Motivator | Hygiene Factor |
|---|---|

*Figure 4-2*

The comparison reveals distinct priorities across leadership levels, shaped by their proximity to frontline staff. Directors and managers show significantly higher concern for salaries and benefits (82%) than senior executives (65%), a 17-percentage-point difference likely reflecting their closer day-to-day involvement with employees and direct exposure to compensation concerns.

This hands-on relationship with staff also appears to drive directors' and managers' greater emphasis on fostering positive coworker relationships (61% versus 52% for senior executives). In contrast, senior executives prioritize recognition more heavily (75%) than their middle-management counterparts (64%), possibly reflecting their strategic focus on organizational culture and talent retention at the enterprise level.

The data suggests that leadership priorities naturally align with role proximity: those closer to individual contributors focus more on tangible benefits and interpersonal dynamics, while those at higher organizational levels emphasize broader recognition strategies.

Functional specialization drives distinct leadership priorities, with cybersecurity leaders showing particularly pronounced differences. CISOs demonstrate a notably stronger commitment to employee development than directors and managers (72% versus 61%), while also emphasizing competitive compensation more than senior executives (75% versus 65%).

These priorities directly reflect the unique pressures of cybersecurity leadership. The rapidly evolving threat landscape demands continuous skill updates and specialized expertise that becomes obsolete within months. This creates an intensely competitive talent market where cybersecurity professionals command premium compensation and expect substantial investment in their professional development.

For CISOs, the dual focus on development and compensation isn't just about retention; it's about maintaining organizational security in an environment where skills gaps can expose companies to catastrophic risks. The data suggests that specialized roles with high stakes and rapid change cycles naturally drive leaders toward more aggressive talent strategies.

In addition to role-based differences, the survey data uncovers intriguing trends when examined by geography, gender, and company size. Further insights from the survey include:

- Geographic differences: U.S. respondents rate the quality of supervisors 12% higher than their non-U.S. counterparts (57% vs. 45%), while non-U.S. leaders place greater emphasis on working conditions, ranking them 16% higher than U.S. respondents (42% vs. 26%).
- Gender differences: Male and female responses are broadly similar, except in two areas: men rank salaries and benefits 9% higher (70% vs. 61%), and are 8% more likely to emphasize expanding employee job responsibilities (26% vs. 18%).
- Company size: Leaders from organizations with fewer than 1,000 employees are likelier to value positive coworker relationships than those from larger firms (59% vs. 49%).

## Responses from Leaders
## of Leading Workplaces

While these general trends provide valuable insights, the responses from executives at top-ranked workplaces offer an even more compelling lens into what truly drives exceptional employee experiences. Thirty-five technology leaders who participated in the survey represent organizations recognized as "Best Places to Work in IT" by Foundry's Computerworld. This prestigious recognition has, for years, set the standard for excellence in IT workplace culture, a distinction many aspire to but few attain.

The annual award honors companies demonstrating a strong, ongoing commitment to key areas such as employee training, career advancement, compensation and benefits, and overall engagement and retention. These 35 leaders have received the award an impressive 103 times, highlighting their long-term commitment to fostering an environment where technology professionals can thrive and succeed.

These Best Places (BP) leaders represent various industries, including healthcare, higher education, insurance, and manufacturing; most are from the United States (91%), with a gender distribution of 74% male and 26% female.

Half are employed at firms with more than 5,000 employees, 38% represent companies with 1,000 to 5,000 employees, and 12% work at organizations with fewer than 1,000 employees. The data from these leaders are shown in **Figure 4-3** and are compared with those of their peers, senior executives.

| What are the five most important areas you focus on to create a great workplace? | | | |
|---|---|---|---|
| Factor | Best Places Leaders | Senior Executives (Excluding Best Places) | Difference |
| Salaries and benefits | 79% | 65% | 14% |
| Quality of supervisors | 64% | 52% | 12% |
| Company policies and procedures | 10% | 8% | 2% |
| Positive coworker relationships | 50% | 52% | -2% |
| Working conditions | 30% | 34% | -4% |
| Employee development | 70% | 68% | 2% |
| Assigning meaningful work to employees | 61% | 71% | -10% |
| Expanding employee job responsibilities | 24% | 23% | 1% |
| Recognition | 59% | 75% | -16% |
| Employee career advancement | 53% | 49% | 4% |

| Motivator | Hygiene Factor |

**Figure 4-3**

## What Sets BP Leaders Apart

When examining how these BP leaders approach building a great workplace, distinct patterns emerge that help explain their sustained success. Compared to their peers, BP leaders stand out in several important ways. They place greater emphasis on supervisor quality (+12%) and salaries (+14%), signaling their belief that a strong workplace begins with fair compensation and highly capable supervisors who drive engagement and enable team members to succeed daily.

By contrast, these same BP leaders give less weight to employee recognition (-16%) and meaningful work (-10%) than other senior executives. While these factors matter, BP leaders see them as outcomes, not starting points. For them, recognition is the byproduct of strong supervisors who help employees grow and succeed at meaningful work. This structured approach creates an enduring culture of excellence where employees feel supported, developed, and empowered.

This perspective also reveals a deeper philosophical difference. BP leaders pursue a more balanced use of hygiene factors and motivators. As shown in **Figure 4-4**, they maintain only a 6-point gap in prioritizing these factors, compared with a 16-point gap among their peers. One six-time Best Places honoree captured this mindset well when asked which elements mattered most: "All of them equally."

| What are the five most important areas you focus on to create a great workplace? | | |
|---|---|---|
| | **Best Places Leaders** | **Senior Executives (Excluding Best Places)** |
| Hygiene Factors | 47% | 42% |
| Motivators | 53% | 58% |
| **Hygiene / Motivator Difference** | **6%** | **16%** |

*Figure 4-4*

## Having a Long-Term Perspective on Workplace Culture

This balanced approach isn't just meant for quick results; it's designed for sustainability. Another critical reason to have quality supervisors on your team is to ensure the sustainability of a highly engaged organization. An analysis of a single year's "Best Places to Work in IT" award found that 75% had earned the honor multiple times over the prior nine years, demonstrating enduring workplace excellence. Remarkably, 41% of those repeat honorees had done so under different senior leaders, underscoring that creating a culture prioritizing employees will lead to a highly resilient and engaged organization that will endure long after senior leaders leave.

By instilling the values of a people-first mindset at all levels of supervision, organizations can create a self-sustaining environment that supports the technology team, strengthens the firm, and, most importantly, champions the people who make it all happen.

## CHAPTER 7

# Impacting the Team: How Do They Do It?

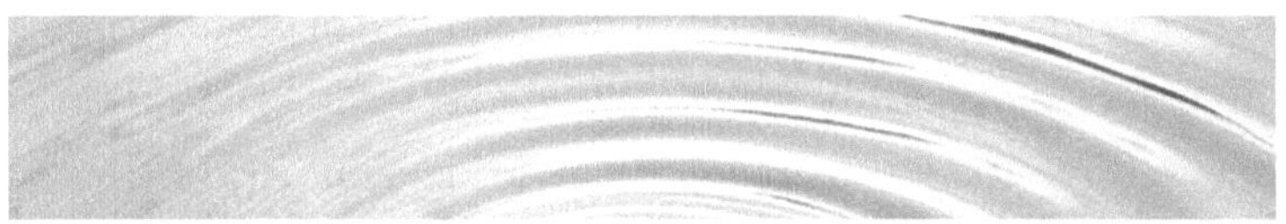

These leaders don't just manage teams; they inspire them. Through thoughtful listening, transparent communication, and the courage to act, they've built environments where employees feel valued, motivated, and empowered to thrive.

## STEVE ZERBY
### Firm: Owens Corning

What does it take to win Best Places to Work in IT eight times and rank number one among large companies? Consider the example of Steve Zerby, who proved it wasn't about having the biggest staff or biggest budget. It was about something far more fundamental: putting people first.

Zerby spent over two decades at Owens Corning, a global leader in building and construction materials, including ten years as Senior Vice President and Chief Information Officer. Throughout his tenure, he prioritized building and sustaining a workplace where employees could thrive. While the CIO role typically carries expectations around technical excellence and organizational leadership, Zerby believes one critical dimension is often undervalued: being a leader of people.

Zerby observes, "There's one lesson I've learned over the decade in this job: to be a great CIO, you must be a risk manager first, a talent manager second, and a tech manager third. If you get any of those out of order, you will regret it." His approach underscores a key truth: technology leadership isn't just about managing systems; it's about managing risk, nurturing talent, and creating a culture where people feel seen, supported, and motivated to contribute their best.

## Creating A Great Place to Work Begins with Hiring

This talent-focused mindset shapes every aspect of Zerby's leadership, beginning with recruitment. The Owens Corning IT organization's brand is "low ego, no ego." Even highly skilled technologists are passed over if their egos don't align with the team culture. This expectation is made clear to every candidate. While some have questioned whether such candor might deter strong applicants, Zerby embraces the transparency: "It paints a clear picture of what life is like here. Some people will run at that, and others will run away. I'm confident the brochure will match the experience when they arrive." This upfront honesty not only strengthens the team culture but also improves retention. Zerby says, "If you have a strong brand that causes people to run away rather than join you, that's just attrition that never happened."

## Attracting and Retaining Talent through Meaningful Work

This hiring philosophy also shapes how the team presents opportunities. They emphasize that while Owens Corning is a large company, its IT team is small, allowing new employees to take on meaningful roles early in their careers. They're not searching for candidates who want to be among hundreds doing similar work, but for those eager for significant responsibility from the beginning of their careers. The team's size makes this possible and speeds up employee growth.

New hires are often the first to assume a new role in the IT organization. Zerby explains, "Some people love that, and some say they love that but are scared to death, and they don't come to work here." Being clear about the culture and expectations is crucial for attracting and keeping the right early- and mid-career talent.

## Developing Talent by Expanding Responsibilities

Being a leader requires balancing the needs of a business and its employees, which can be challenging. Firms want consistency, results, and no surprises. Whether it involves budgets, project deadlines, or IT staff, the organizations we serve prefer not to worry about the technology operations. Likewise, employees want engaging assignments, updated skills, career advancement, and the resulting salary increases. Attempting to satisfy both of these may cause a conflict that, if not handled carefully, could lead to disappointment for the business, the employee, or both.

Zerby believes you can create employee developmental opportunities while satisfying your internal stakeholders. He recognizes that his business counterparts don't want to retrain a new IT person who supports their area every two years because the incumbent has left the role for a new opportunity in the company. Zerby believes many firms follow a formulaic approach to structuring the IT organization by putting people into standardized roles like infrastructure, business applications, and software development. He takes a different approach by mixing and matching the IT organization to meet the needs of the business while also addressing the development of top talent. He notes, "I believe this, with every ounce that I have in me, that you need to build jobs around people instead of putting people in jobs."

A concrete example that illustrates this approach is an employee in their mid-twenties who was considered a high-performing talent. As a result, this employee was sent to Europe for a two-year assignment to lead an extensive program for the company. Upon their return, they were given a technology assignment but eventually left the company for a role in consulting. When Zerby later became CIO, he recognized the missed opportunity and recruited this person back to the company. He pledged to give this individual various experiences by building positions and, therefore, more responsibility.

After returning to the company, their initial assignment was to serve as the director of IT strategy. Over time, the role was expanded to include enterprise project management. Next, they took on additional responsibilities as a divisional CIO for one of the businesses. Finally, digital IT for market-facing initiatives was added to their duties. These extra responsibilities helped expand this employee's depth of experience while maintaining continuity in the relationships built between the IT organization and the business areas they support.

This approach has been successful at Owens Corning because business stakeholders have the comfort of having the same person on their leadership team while enabling this individual to have a wide range of growth opportunities. As a result, the credibility of the IT organization is elevated through a stable and trusted relationship with your business colleagues while enabling employees to grow their careers without leaving your company.

## Daily Focus on Retaining Talent

Building jobs around people reflects Zerby's core belief about retention: successful technology leaders dedicate more time to talent and people than to technology. Conversely, he believes leaders who dread prioritizing their team members and understanding what motivates them in their work and personal lives are doing a disservice to their employees. These leaders are also doing a disservice to the firm that employs them by creating a work environment that can be viewed externally as uncaring and disinterested in their staff.

Zerby's response to this challenge is a practical approach to creating a highly resilient and engaged workplace. He believes you need to "get up every morning and work as hard to retain the people you have as you're going to have to work to replace them when they leave." In doing this, Zerby believes you won't have a retention problem, and therefore, "you don't have an attraction problem" either. He recognizes that many leaders approach their time with employees by focusing on periodic discussions about compensation, promotions, and objective setting, or spending significant amounts of time with them, which may not be sustainable given the demands of being a technology leader. Zerby doesn't subscribe to either of these approaches. He believes "you can invest minutes a day with employees in key moments and build significant emotional capital with those around you that causes them to stay."

This approach has resulted in a low attrition rate among the IT team at Owens Corning. Zerby acknowledges that when you have a significant attrition rate, it has an accelerated impact on the organization. First, you spend time and energy on recruiting. Second, if you lose ten percent of your team, you need to have the other ninety percent absorb the work, which takes a toll on the organization. Finally, when the team sees others leave the organization, it lowers the psychological barriers for others to quit because it "doesn't seem quite as scary."

### A Legacy of People-First Leadership

By understanding and preventing this domino effect, Steve Zerby has cultivated a leading workplace culture that is recognized inside and outside Owens Corning. Focusing on his most critical asset, his team members, Zerby's impact will endure beyond his time at Owens Corning, establishing a benchmark for future leaders in the IT organization.

## GRAEME THOMPSON
### Firm: Informatica

In the heart of Silicon Valley, where tech talent can select from many well-known industry leaders, Graeme Thompson faced a challenge that would define his tenure as CIO at Informatica. The company was shifting from a traditional license-based model to subscription and cloud services, a change that would succeed or fail based on its IT team's ability to execute. But first, he needed to understand why they should care.

Thompson, a CIO who has spent most of his career in the technology industry, would need to draw on all his experience to meet this challenge. At Informatica, a software provider that helps organizations strategically manage their data in the cloud, he has led an IT organization that Computerworld has repeatedly recognized as an outstanding workplace. His success has resulted from his focus on connecting his team to the company's mission and providing them with the opportunities and skills to enable the firm to achieve its goals.

## Listen to Learn About the Culture

When Thompson joined Informatica as the CIO, he conducted a listening tour of the company with colleagues from marketing, sales, customer success, and product development. A key takeaway from this exercise was that he learned that the people in the company cared about their coworkers and were genuinely interested in helping each other succeed. This supportive culture is strengthened by the company's mission-driven environment, in which employee goals align with the organization's objectives. Salespeople understand how important it is that the customer adopts and gets value from their products, so they want to support the customer success function in performing their work. The finance department knows that it is essential to be able to record and report revenue accurately. As a result, they will help the sales team structure deals appropriately to ensure that reporting is timely and accurate and that salespeople are compensated appropriately. When one group faces a problem, others within the company will help solve the issue. Thompson noted that when you're trying to solve a problem with another function at Informatica, it "feels like they're in the boat with you."

In other companies where Thompson worked, employees focused on the success of their function rather than that of the company. In working environments like that, he observed that individual objectives might not be aligned, leading to conflicts between different groups within the same firm. After hearing from his Informatica business counterparts, Thompson knew that the company's supportive and mission-based culture was critical to its success and, therefore, the success of his IT team.

Once he understood Informatica's culture, Thompson realized that one of his first tasks was to align his team with the company's mission. He realized that his IT team members passed other tech companies like Netflix, Microsoft, and LinkedIn on their way to the San Francisco Bay Area office, and he needed to ensure he provided them with a compelling reason to choose Informatica over other local employers.

## Connecting Employees to the Company's Mission

Leaders at Informatica understand that employees are interested in more than a job and a paycheck. They know that good employees are mission-driven and want to be a part of something bigger than themselves. These employees want to be able to connect with the firm's mission, and they want to know that what they are doing has lasting value. Thompson notes, "We spend so much time at work that what we do has to be worth our time. Why else do it?" He goes further to say that people in the technology field can choose where they spend their time and will ask themselves whether working for their company is worth the effort and stress to be successful. As a result, companies want to be able to articulate the employee value proposition and quantify the benefits of working for their firm. Thompson believes that for the companies that understand this, "it's their secret sauce that they just didn't know was there." He further notes that "being intentional and open about what's great about your company is one of the untapped resources we all have as leaders to attract and retain our great people."

When the company went private in 2015 and revised its business model, Thompson knew he had to articulate Informatica's "secret sauce" to connect his team to its revised goals and retain his most talented employees. The new owners were focused on transforming the company's sales from license and maintenance-based to subscription and cloud-based. The leadership challenge was to translate what that meant to the IT team so that they could help change legacy processes and systems and help the company generate revenue in a new way. Thompson understood that IT would be critical in the company's future and that his job was to guide his team along the transformation journey.

The business objective was to increase cloud-based subscription revenue tenfold within four to five years. While everyone in the organization understood that IT wasn't leading the transformation to the cloud, it couldn't be enabled without their support. This meant that all go-to-market, customer-facing applications and the enterprise resource planning (ERP) system had to be replaced to support the new subscription-based revenue model. Thompson made it clear that the team would be critical to meeting this objective; therefore, what they did was important and aligned with the company's mission.

Informatica successfully transformed its business due to the contributions made by all of the firm's functions, including IT. As new employees join the IT team, they hear this story and understand how their role is critical to Informatica's ongoing mission.

### Maintaining the Employee-Mission Connection

Connecting employees to the company's mission does not stop once a business goal is achieved; it is vital to continuously direct your team toward what truly matters in fulfilling the firm's mission. Thompson learned from a customer that the most challenging part of being a leader was getting the brightest people to focus on the right things. Smart people often want to work on the most interesting problem, which may not be profitable. He works to ensure that all his team members know that while a specific work assignment may not be the most appealing, it does connect with the company's overall success.

Thompson explains the role of the IT function in achieving the company's objectives by comparing it to a baseball team's third catcher. He notes, "No one wants to be the third catcher on the team that wins the World Series. The team needs you, but they hope they don't have to rely on you. That's not IT at Informatica. We are on the critical path to being successful. The company will not succeed because of IT, but it can't be done without us, which is incredibly powerful." This perspective inspires Informatica employees to view their roles as vital in reaching the company's objectives and motivates them to develop and apply their skills to help the firm succeed.

One way the company reinforces the importance of its mission is to continually connect employees with how their products help customers. As Thompson notes, while many of Informatica's customers benefit from better data, resulting in improved financial performance, these outcomes don't necessarily make your employees proud to work for the company. He knows that when an employee's family or friends ask what their company does, they probably don't say they are helping a bank make more money. Informatica employees feel good when they wake up in the morning and know that their work helps to improve the lives of others.

During the COVID pandemic, local and state governments used Informatica's software to contact residents and inform them of testing availability and safety protocols. In addition, pharmaceutical companies have used Informatica's technology to help progress their vaccine through clinical testing. Employees are motivated to help their employer when they know their contribution can help save lives across the globe. They proudly tell their friends and families what they do for a living and how their work helps their company make a difference. As a technology leader, it is essential to continually find opportunities to make this connection to reinforce each employee's contribution to the greater good.

## Providing Meaningful Work and the Associated Skills

When team members understand and connect with the company's mission, they can see their work as meaningful. Thompson notes, "For some employees, that's the fire that burns inside and gets them going daily. It may not be for everyone on the team, but it is essential for those who want to innovate and make a difference." Many software industry firms have struggled to transition to a subscription and cloud revenue model. Informatica IT employees know they have achieved something that will be respected in the industry as ground-breaking, thereby creating a personal "halo effect" related to their contribution to this accomplishment.

Although the primary role of the Informatica IT team is to support the overall business operations, some members have the opportunity to collaborate with the product management group in Research and Development (R&D) to develop new versions of the company's software. These are sought-after roles that enable them to work with and learn from some of the company's top technical talent. They also serve as the initial users of a new product and welcome the chance to be directly involved in improving it for the end customer. The IT team members value these assignments and take pride in knowing their contributions have positively influenced the company's technology.

Like most companies, Informatica's IT portfolio includes modern and legacy systems. Thompson understands that his team members prefer to work on newer technologies and develop the associated skills. When they do so, they are proud to be working on an industry-leading technology platform rather than maintaining legacy systems. A modern tech stack can benefit both the company and the employees supporting it. For the company, a new system can help improve the firm's operations. The employees who implement and support the technology develop new skills, making them more valuable to the company and, therefore, more likely to continue working for the firm.

Regardless of the industry, technology employees want the skills that provide value to their employer. At Informatica, the company offers online learning via Udemy, LinkedIn Learning, and Coursera because its leadership understands that investing in its employees is also an investment in the company. They also know that while training opportunities are made available to employees, they must ensure they have time to participate in the course; otherwise, they'll never be able to develop their skills.

IT leaders at the company address this during the annual objective-setting process by identifying a tailored curriculum to match the employees' career aspirations. With the support of their supervisor, an employee selects a curriculum, and then they both agree on the schedule to complete the training. The supervisor then acknowledges that the employee will have time to complete the training. This approach makes the employee and their supervisor accountable for executing the development plan and reinforces to the employee that the company supports their ongoing career development. Since Informatica will fund and benefit from the training, the employee must be accountable for creating and executing the development plan. This can help team members think about how to develop their careers and engage their supervisor to assist them.

Graeme Thompson's leadership at Informatica demonstrates the lasting impact of a tech leader who champions purpose and people. By consistently aligning his team's work with the firm's broader mission, he ensures that employees understand what they are doing and why it matters. Thompson emphasizes giving team members the skills, opportunities, and support they need to contribute meaningfully to the organization's success. This approach has fostered a workplace culture that motivates employees to grow their careers with the company while positioning Informatica as a destination employer for top technology talent. As a result of this commitment, Informatica has been recognized as one of Computerworld's "Best Places to Work in IT" for 12 consecutive years, a distinction that reflects both organizational excellence and a sustained focus on building a high-performing, engaged technology team.

# HANS KELLER

Firm: **Erickson Senior Living**

When Hans Keller looks at the IT organization at Erickson Senior Living, he doesn't just see systems and projects; he sees a team enabling a network of senior living communities and the residents and employees who depend on them daily. As Senior Vice President and CIO, his work sits at the intersection of technology, people, and enterprise strategy. His focus is on developing people and leaders to keep pace with rapid changes in technology, the senior living industry, and Erickson's evolving business priorities.

Keller's belief is simple: technology will continue to evolve, but long-term success depends on peoples' skills, character, and ability to lead through change. Over time, this philosophy has helped Erickson's IT organization earn repeated recognition as a top workplace for technology professionals, including national Best Places to Work in IT honors. More importantly to Keller, it has fostered a culture where people come to grow their careers, not just to do a job.

## Developing People for What's Next

Keller and his leadership team base their approach on a few simple ideas: stay curious, develop strong and adaptable skills, and work as part of something bigger than yourself. These concepts are reflected in how the organization nurtures its people.

Encouraging curiosity involves looking beyond your current role. A skilled developer might spend time with infrastructure, cybersecurity, or data teams. Similarly, a service desk analyst could join a project team working on new resident- or staff-facing capabilities. These experiences help people see how the entire IT ecosystem fits together and understand how their work aligns with Erickson's broader mission and strategic priorities.

Competency is viewed as a dynamic and intentionally managed aspect. Erickson IT employs individual development plans (IDPs) and a skills-focused development approach rather than relying solely on job titles and traditional career ladders. Leaders discuss with team members the skills needed for the next wave of work, such as cloud, data, cybersecurity, automation, AI, vendor management, and especially change leadership, and assist them in charting a course to develop those capabilities.

Instead of a rigid "ladder," Keller describes this as a career lattice. People move across domains as well as upward, gaining experience in operations, applications, support, and project or product work. Development conversations directly align with Erickson's enterprise strategy and the company's focus on people, culture, and excellence. The question is not just "what do you want to do next?" but "what skills and experiences will make you most effective as our strategy advances?"

Community connects those ideas. Within IT, that means leaders engaging with their teams, focusing on communication, coaching, and cross-functional collaboration. Outside of the IT organization, it shows up in how the organization engages with the wider community and the senior living mission, reminding everyone that their work makes a real difference for residents and staff each day.

### Growing Leaders from Within

One outcome of this philosophy is a strong record of internal mobility. Several of Erickson's current IT leaders started their careers on the service desk and now lead critical functions such as enterprise applications, network, and telecom. Their career journeys send a clear message to new team members: if you do perform well, continue to learn, and live the team's values, you can advance your career without leaving the organization.

Keller's own path looks different but reinforces the same point. Before joining Erickson, he served in the U.S. Air Force and held technology roles at organizations such as the National Aquarium. Those experiences across different industries shaped his view that effective IT leadership is as much about understanding people and business context as it is about technology. He now encourages his team to intentionally seek breadth of different experiences, different roles, customer groups, and perspectives, as part of their development.

This mindset also appears in how Erickson IT manages change. IT partners closely with a Change Excellence function to plan, communicate, and support major initiatives affecting residents and employees. Team members are not just asked to implement new systems; they are expected to help others adopt and thrive with those changes. For Keller, that is another dimension of leadership development: teaching people to navigate ambiguity, influence across functions, and support others through change.

## Hiring for Character, Expecting High Performance

The same philosophy that shapes internal development also guides recruiting. Attracting talent to work in senior living communities can be challenging when competing with large technology companies, federal agencies, and consulting firms. Keller's answer is not to outspend those employers on salary, but to offer something different: meaningful work, strong culture, and real opportunities to develop skills tied to enterprise priorities.

When interviewing prospective employees, Erickson's IT leaders look for a combination of competence and character.

Technical skills matter; the work is complex and mission-critical. But Keller insists that how someone treats colleagues, customers, and partners is just as important as what they know on day one.

One candidate, for example, clearly had exceptional technical skills. When asked how they saw themselves fitting into the team six months after hire, the candidate replied, "Everybody will think I'm an ass. That's because I'm usually right, and people don't like it when they're not right, and I am." The interviewers thanked the candidate and wished them well elsewhere. For Keller, this story is a simple illustration: the organization will invest in building skills, but it will not compromise on how people show up as teammates and leaders.

Diversity is also an intentional part of the equation. The IT leadership team includes a significant number of women and leaders from different backgrounds and career paths. Erickson's interview process is designed so candidates can see themselves reflected in the organization and talk with people whose career journeys might resemble the one they hope to build.

### Listening to Learn as a Leadership Discipline

Keller often describes listening as one of the most underused tools in a technology leader's toolkit. In his experience, IT organizations are wired to solve problems quickly, sometimes so quickly that leaders jump to solutions before they fully understand what their teams are saying.

At Erickson Senior Living, the IT organization intentionally creates space for people to be heard and to grow. Employees are encouraged to learn or explore something new that may not be directly tied to their current job duties, and leaders are expected to listen for where those interests intersect with the team's mission. Keller notes that if you say your people are your most important asset, you have to be willing to spend your time with them; otherwise, your actions will tell a different story.

To make listening a consistent practice rather than a one-time event, Erickson IT has built several practices into how it operates: team "pulse" conversations, retrospectives after major initiatives, and periodic skip-level discussions where employees can speak directly with senior leaders. In these forums, teams are encouraged to talk candidly about what is working, what is getting in their way, and what they would keep, change, or stop if they could. Every team member is invited into the process so that everyone has a voice in shaping how the organization works.

The insights from these conversations inform how Erickson IT runs its portfolio of work. Understanding where teams are struggling or where processes are breaking down gives leaders better information when they prioritize initiatives, adjust staffing, or refine how changes are rolled out and supported.

Keller observes that as IT professionals, we often prioritize solving problems quickly, sometimes even overnight. He emphasizes the importance of listening to employees, whether they are discussing their developmental goals or the challenges they face in completing their work. "Sometimes you need to listen more than once to understand the whole story fully," he says. It is essential to "allow ourselves the grace to spend time listening without immediately trying to solve someone else's problem, especially regarding our team members."

### Community, Purpose, and Retention

Community has long been a defining part of Erickson's culture, and IT is no exception.

Keller exemplifies this by serving on the advisory board for Towson University and as Vice President of the Mid-Atlantic CIO

Forum, where he also leads the scholarship committee. He also serves on the board of the Baltimore Area Council of Scouting America, and each spring and fall, he and other volunteers from the IT team spend time at Broad Creek Scout Reservation helping to prepare the camp for upcoming seasons of Scouts.

Within the IT organization, employees have organized food drives and friendly competitions to support food-insecure families, volunteered with the Maryland Food Bank to prepare meals, and participated in efforts to build beds for children who do not have one. These efforts are not centrally mandated programs with executive sponsors and dashboards. Instead, leaders create space and remove obstacles, and team members step up to organize and participate.

Keller does not need to lead every initiative personally, but he makes a point of being present, volunteering alongside his team, and acknowledging their efforts. He believes that when employees feel supported in doing work that matters beyond their job description, it strengthens connection, loyalty, and pride in the organization. In a competitive talent market, that sense of purpose can be as important as compensation.

## Principle-Based Leadership in a Time of Change

Keller's approach challenges a familiar assumption in technology leadership: that success is primarily about choosing the right platforms, tools, and architectures. For him, those decisions matter, but they rest on a foundation of core principles about how people are hired, developed, listened to, and trusted, and how their work aligns with the organization's strategy.

By prioritizing character alongside competence, encouraging people to stay curious and build broad, durable skills, and anchoring the work in community and purpose, Keller has helped build an environment where people stay, grow, and step into larger roles. Many leaders have advanced from frontline roles to positions of significant responsibility. Teams that once focused mainly on keeping the lights on now help lead complex initiatives in areas like digital transformation, data, cybersecurity, and enterprise change, because their leaders invested in their growth.

In a field often dominated by talk of the latest technology trends, Keller's message is deliberately simple: "If we don't take care of our people, nothing else matters." Technology stacks will change. Business models will evolve. But the work of leading people, listening, developing, and creating an environment where they can do their best work, remains constant.

For technology leaders trying to build resilient, high-performing organizations in a time of relentless change, that may be the most durable advantage they can create.

## CHAPTER 8

# Impact Story: "If you focus on the people, the technology will follow."

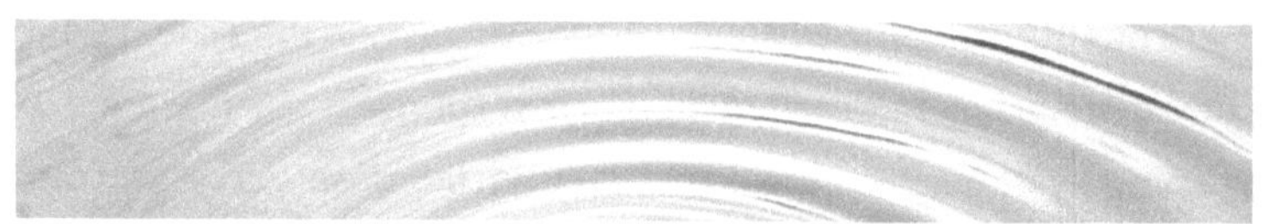

This philosophy would prove essential for a CIO who faced a seemingly impossible task: developing technology capabilities faster than her bank's ambitious growth plans, with an underperforming team that had to evolve quickly to meet the needs of a growing business.

The following is a story about one technology leader who, after more than 16 years in the utility industry, changed employers to become CIO in banking despite having no prior experience. The year was 2012, and the bank she joined had just begun a growth journey that would increase the assets it managed by 12-fold over a decade. Her approach to transforming the organization required her to employ Herzberg's Two Factor theory of hygiene factors and motivators to create a team that would eventually be recognized as one of ComputerWorld's "Best Places to Work in IT."

## Stage 1:
## Recognizing the Need to Change

The CIO faced a daunting challenge: develop technology, talent, and systems capabilities faster than the bank's business growth. Despite her observations and feedback from key stakeholders, she identified that the first hurdle was an IT organization that believed it was performing well. She realized the team didn't know what good looked like, and they thought they were good. It was clear that they needed to recognize the need for improvement to grow as a team. With steadfast determination, she set out to transform the team's mindset and performance.

The CIO immediately engaged the team by asking them what they believed they excelled at. Their response, "We are good at keeping systems running," was met with a dose of reality from the CIO. She had only been in the role for a month, and there had already been four nights of programs failing and multiple out-of-service ATMs. "I am new to banking systems," she admitted, "but that doesn't sound good to me." A staff member responded, "Well, we are good at fixing things when they break." The CIO, with a clear vision, seized the opportunity to steer the team's focus, stating, "We don't want to be good at fixing things when they break; we want to be good at keeping things from breaking."

One example of how the team learned the importance of preventing system outages occurred following a significant infrastructure issue impacting their customers. The head of operations was asked to share stories with the IT team about the impact of the outage on the bank's customers. These stories included a person who could not obtain a loan and couldn't purchase an engagement ring. Another example was an individual impacted by the outage, which caused him to miss a bus that would take him on a cruise.

These stories drove home the point that they were not supporting servers; they were helping customers who needed to perform banking transactions. This enabled the IT team to better understand their impact on the bank's business in ways they had never considered. The team realized, "We periodically need to remind ourselves about what the technology we support means to our end customers."

## Focus on Coworker Relationships

Next, everyone in the organization was asked to rate their team and the IT department. Each team rated itself as a 3 but the department as a 1. The CIO asked them, "How can that be? If all of our teams are good, how can we be bad as a group?" It was apparent that there were silos within the department. She told them that their customers didn't care if there was a problem with the network or a database; they just wanted the system to work. She said that "we should all care about our weakest team" and collectively work together to support them.

As an exercise, the team was asked to select two pictures. One depicted what life at work was like now, and the other was what they wanted it to be like in two years. To visualize the current state, they selected pictures of cars running into each other and the future state of a highway where traffic is flowing smoothly. The CIO knew that once they began to visualize the chaos of the current state, they could start the improvement process and build credibility with their customers.

## Stage 2:
## Rebuilding the Organizational Foundation

With the team now able to see their current problems and their desired future state, the CIO knew they were prepared for the challenging work ahead. But awareness alone wouldn't be enough; she needed to tackle the barriers that had caused these issues in the first place.

To improve the current situation and reconnect with the bank's mission, the IT team adopted the mantra that "we are the bank, and we need to understand banking." They knew that they needed to shift the organization's mindset from supporting systems to supporting bankers. They realized they were not performing at the level the bank required and had to improve quickly. That meant the leaders who were not performing would need to leave the organization. During her first few months, 30% of the IT leaders were separated from the bank.

The CIO was asked how she could quickly conclude that so many had to leave. She had "triangulated it from conversations with human resources, business partners, and some of the best IT leaders on the team." They all identified the same list of people to leave the firm. The IT team needed to see her act quickly so the organization could embark on its improvement journey.

It took the team about two years to finally align on a common view of their performance. There was "heavy lifting, changing out of leaders, setting expectations, and messaging repeatedly." The team also began to focus on execution. The leadership team read the book "Execution: The Discipline of Getting Things Done" by David Bossidy and Ram Charan to help them understand what it takes to execute successfully as a unified organization. They then began to prove their worth by targeting specific execution areas for improvement and thereby creating small wins as a team.

## Celebrating Accomplishments and Supporting Development

In conjunction with successful execution, the CIO knew she needed to recognize the teams' efforts, big or small. She believes there is no "secret sauce" to recognition other than just doing it. If she heard that a team member did something well, she would email them to thank them. In addition, if a member of her staff sent an email thanking a team member, they would also copy her. The CIO would reach out to that individual and thank them in person because she knew that it was essential to take the time to reinforce a culture of recognition.

In addition, if something didn't go well, she would contact the responsible person and let them know she was aware of the problem and share her expectations. The team quickly learned that their CIO was invested in their success and would engage with them in good and bad times. This, in turn, built a sense of trust between her and the rest of the team. They knew that the CIO was invested in their performance and that she wanted them to grow and develop professionally. She would help them review what went wrong and other ways they could have handled the situation. She would tell them, "If you come to me before you have a problem, I can help. If you come to me after the problem occurs, I must be a judge." As a result, the team realized that she wouldn't fire them for a mistake; she would give them feedback.

## Stage 3:
## Activating the Transformed IT Team

After two years of hard work, the foundation was finally secure. The team shared a common understanding of performance, trusted leadership, and had a proven track record of execution. Now the CIO could focus on unlocking their full potential.

## Empowering Employees by
## Expanding Job Responsibilities

IT leadership at the bank recognized that they weren't developing first-line managers. There was a gap between strong individual contributors and first-line management. The managers felt that they were taking on far more than they should. To address this, they developed an informal leader training session, reinforcing that team members didn't need to wait for a manager to tell them what they could or couldn't do. While it was clear that each person knew their job responsibilities, they were never told that they were expected to lead in their roles. To grow the power of the organization with the existing team members, the CIO and the senior IT leadership team made sure that they knew they were trusted and empowered to lead.

As the bank grew, so did the IT team members. With each new acquisition, top performers had opportunities to "step up and take on more responsibility." Newly acquired businesses also provided an infusion of fresh talent. The company's growth also meant expanded responsibilities, teams, budgets, and new project opportunities for IT team members.

## Challenging The Status Quo of the Firm

As the business grew, the firm's norms, policies, and procedures began to be scrutinized. One area that the CIO wanted to challenge was to establish the bank's first internship program. While the bank had never hired interns, she knew this would enable a pipeline of cost-effective talent to add to the organization. She convinced the bank to let her "pilot" the program in the IT department. The program's goal was to be known for requiring interns to perform "real" work. The test of whether or not the work was "real" was when an intern could go home and show their parents a website that they updated or discuss a software program they had developed and was being used by the bank.

After the first two years, the program was so successful that the bank adopted a company-wide internship program. Approximately 50% - 60% of all the IT interns are offered employment at the bank, which serves as the "feeder stock" for getting fresh perspectives into the organization.

Driving change at the bank and within the team was a way to reinforce that the team was maturing and gaining the trust of the organization's senior leadership. A simple but meaningful change was the introduction of "Jeans Friday." While this has since become commonplace, it was considered somewhat radical for the culture of the broader organization. This was used to "reinforce the winning team concept" and to demonstrate the CIO's willingness to try something new. Jeans Friday, in turn, helped shape the IT organization's new culture as a driver of change. Eventually, the bank adopted the IT-led initiatives of casual dress and internships.

# Stage 4:
# Transformation Realized

These early successes in challenging the status quo were more than just policy changes; they proved that the IT team had become confident agents of change. The actual test of this transformation would be when they had to challenge company policies and executive leadership.

With the success of the internship program and "Jeans Fridays," the team began to feel empowered to promote more of their ideas. This was manifested at the beginning of the pandemic in 2020. Just days before deciding to send employees home to work remotely, the IT team told the CIO that they needed to stress test the systems to ensure that they would support the anticipated impact of many remote workers. When she reviewed the idea with the executive team, they expressed concern that it would "scare the rest of the organization and that they will think that we're going to shut down."

When this concern was shared with the IT team, they told her they needed to perform this test to validate that they could support the bank remotely. They urged her to return to the executive team and insisted that they conduct this exercise. They agreed to call it a "business continuity disaster recovery test" to minimize the concerns of the other employees at the bank. The entire IT team worked remotely on Thursday and Friday before the eventual decision to send everyone home was made the following Monday. This enabled the IT team to verify that they could access the systems required to support the rest of the bank.

This experience proved that the team's culture had changed, so they were willing to challenge executive leadership to do the next right thing. They demonstrated that they understood the potential impact that the pandemic could have on the ongoing operation of the bank, and they knew that they had the responsibility to highlight the need to be prepared for the impending challenge. As a result, the bank successfully transitioned to remote work with IT support. This experience also helped to strengthen the trust between the CIO and the IT organization. In the words of the CIO, "There are some battles that your team needs to see you fight."

## The Journey to a High-Performing, Award-Winning Workplace

The transformation of a small, siloed group of IT teams into a large, cohesive, high-performing organization did not occur overnight. The CIO in this story showed patience and persistence by focusing on the people, their mindset, development, and empowerment, as the technology capabilities had indeed followed, supporting the bank's remarkable growth journey. By aligning her team's work with the bank's mission and consistently choosing to do the next right thing, she helped build a stronger IT organization that supported the bank's remarkable 12-fold growth and a workplace recognized as one of the best in IT.

## CHAPTER 9

# Impacting the Team: Communicating to Connect

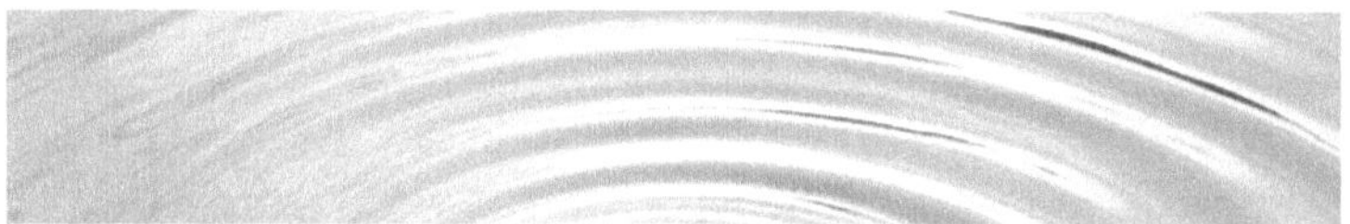

Imagine that you have decided to move the firm's long-established enterprise resource planning (ERP) system from Oracle to SAP. When communicating this decision, you should expect many questions and concerns. Art, your database administrator, may immediately focus on the underlying database technology and wonder if that will also change. Your senior Oracle business analyst, Cathy, may immediately wonder if her job is secure and ask if she should begin looking for a new employer. Others on the team will ask if they will be included in the project and whether they'll receive training. One business unit may push to be an early adopter, while others may entirely resist the change in business systems to avoid disruption. Meanwhile, senior executives like the CEO and CFO will care about cost, timeline, and return on investment.

This single announcement demonstrates a fundamental challenge: a wide range of reactions shaped by each stakeholder's priorities and concerns. Understanding and navigating these varied perspectives is a core leadership challenge for IT leaders. You must simultaneously engage technical experts and business stakeholders with differing priorities and expertise. A message that's clear to engineers may confuse board members. Tailoring communication for business leaders may leave technical teams feeling uncertain. Striking the right balance is a critical leadership skill.

The stakes are particularly high in technology organizations, where your decisions can impact the lives and livelihoods of your team members and colleagues. That is why effectively communicating and connecting with the people you work with is essential. The following leadership lesson from Dick Daniels demonstrates the unique challenges of communicating as a technology leader and the importance of the content and how it is delivered.

## Communicate to Connect: Dick Daniels

One technology leader who knows how to communicate effectively is Dick Daniels, a CIO Hall of Fame inductee. During each of the five years he served as Executive Vice President and CIO of the health care provider Kaiser Permanente, Foundry's Computerworld recognized the IT organization as one of the best places to work in IT.

Daniels recognizes that leading IT personnel differs from leading employees in other business disciplines. He sees parallels between the specialization found in the medical and IT professions. Medicine has podiatrists, cardiologists, hematologists, and dermatologists. Likewise, IT specializations exist, such as software engineers, cybersecurity analysts, telecommunications technicians, and database administrators. These diverse specialists must work together to create a comprehensive solution for the firm they support. Daniels knows that effectively leading a technology team with diverse skills requires understanding the technology *and* aligning the organization through clear communication of the mission, vision, and strategy. Conversely, ineffective or, worse yet, no communication can cause confusion and conflicts across the organization.

## Communicating with an IT Audience

Speaking to IT employees can be a challenge. You need to articulate broad concepts to the organization as a whole while also providing the details for your audience of specialists who want to understand the specifics. Daniels uses the analogy that you can't write software by being too broad. As a developer, you need to be very specific about everything that the code needs to do. Similarly, IT specialists will expect a level of detail before they feel you have effectively conveyed your message.

As a technology leader, your organization isn't the only audience you need to communicate with. In your role, you need to speak with business partners, senior leaders, and potentially external parties. Each of these groups of constituents may need to understand the same underlying message, although it will need to be tailored to the specific audience. To do this, Daniels believes a framework should be developed to convey key messages so that all of the leaders within your team can reuse the core information.

In addition to tailoring your content for each audience, effective communication requires repetition to reinforce key points and ensure the organization is on the same page. Equally important, your message should consistently connect to the firm's mission. By adapting, repeating, and connecting your communications to those listening, each person should walk away understanding what you're saying, how it impacts them directly, and why it's vital to the firm's overall goals.

## Adapt Your Communication for Your Audience

The starting point for developing a critical communications framework begins with your team of direct reports. Everyone must be aligned with the organization's mission, vision, strategies, and plans. To do this, Daniels would convene his team of leaders annually to craft the plans for the coming year and the associated communications. During those sessions, he encouraged each team member to participate in the discussion and offer individual opinions, objections, and support. He views this as essential to getting the entire team aligned on the plans they develop and the messaging that they will deliver. The team must speak with one voice. They don't need to use the exact words, but must convey critical messages to their audience using a common framework. When executed effectively, the communication will align with the business strategy and resonate with the IT team and business partners.

Daniels has employed the people, process, and technology framework for communicating and has added a fourth element for financials. He has used this framework to represent IT initiatives and their alignment with the business's strategy. An example of this framework is as follows:

- People: Invest in our people
- Process: Execute with discipline
- Technology: Implement rock-solid solutions
- Financials: Deliver a competitive cost structure

Once you set the framework, you can reuse it for several years. The next step requires developing specific tactics for each area throughout the year. Daniels forms four groups from his leadership team to create the individual elements for the framework. Once completed, the groups review their ideas with the leadership team to obtain buy-in. This way, every team member can contribute and express their thoughts on the organization's priorities.

Daniels emphasizes, "When we walk out of the room, we are all committed to what we have developed as a team." Every staff member is aware that they should be united in their support. Therefore, if they have concerns about something, they should speak up before it is shared with IT and business personnel. Once the framework for that year is completed, it is time to begin using it to communicate with all of the organization's constituents. As Daniels notes, "Whatever words we produced as a team were the sheet of music we were all singing from for an entire year."

## Repetition, Repetition, Repetition

Once you prepare the message, ensure all leaders on the team deliver it consistently and repeatedly. Every opportunity to speak in front of an audience should also be an occasion to share the strategies and plans of the organization using the established framework. As Daniels notes, "You can't tell people something once and think they understand it. It's important to convey your strategy every opportunity that you have." You may need to put your key messages on mouse pads, screen savers, shirts, or coffee mugs to ensure they permeate the organization.

One way Daniels measured the effectiveness of their communications was through the annual culture survey question: "Do you understand how your job contributes to the organization?"

He knew that if his team didn't consistently share a key message, many people would answer "no" to this question. Employees will acknowledge that they heard it if you consistently use a communication framework.

Daniels illustrates the importance of repeating a message through a personal story of his time as a high school marching band member. The band members were exhausted from playing the same songs repeatedly and asked the band director why they couldn't play something new. The director responded, "Something that you are all forgetting is that when we perform our music in a new city, they haven't heard the songs before. You have a different audience each time."

The same principle applies to organizational communication; what feels repetitive to leadership is often the first time many employees truly hear and process the message. While the music was old to the band members, it was new to the audience in each new community they visited. As a leader, you might think everyone has heard the communication because you've repeated it multiple times. However, new members may join the organization, while others may have missed a meeting or did not understand it the first time. Consistently and repeatedly conveying the organization's key messages is essential.

Another technique Daniels has employed is institutionalizing communications by establishing relevant goals with associated metrics and targets. When you report progress on goals, you demonstrate that the organization is making progress, which becomes a source of pride. It helps affirm that the organization is advancing and is the hallmark of a strong organization.

## Connect Your Team to the Mission

Daniels understands that another essential use of effective communication is connecting your team with the organization's mission and purpose. Doing so gives a higher meaning to the day-to-day work and can provide the fuel to keep going when the weight of work becomes heavy.

The Kaiser Permanente IT team members view themselves as part of the larger healthcare team providing patient care. Daniels knows his team members must connect their work to something more meaningful than "I'm just doing desktop support, or I'm working on a particular business application." Helping his team understand the bigger picture and knowing how the technology is being used can be powerful. It's empowering for employees to feel that they are helping to improve the healthcare their firm delivers and, therefore, improve patients' lives.

Daniels believes every IT professional needs the experience of walking in their users' shoes. He made field visits mandatory so team members could see firsthand how users employ the technology. The application team and business analysts would typically be closer to the businesspeople because they would interact with them frequently. That meant regular visits to where users performed their work. However, the team working in the data center or on networks and infrastructure had no reason to do that. Daniels intentionally enabled the infrastructure teams to come out of their workplace to see how the technology they support is being used.

Connecting to the mission of the firm isn't just for your team. With all the challenges and pressures of being a technology leader, you may occasionally need to reconnect to your purpose for working at your firm. Daniels noted that he would periodically remind himself about the importance of his work by stopping at a hospital on the way home and walking through the facility. It reminded him of why the work that he was doing was important. He watched doctors and nurses care for patients and family members worried about their loved ones. He knew he and his team had a role in improving healthcare by supporting the front-line workers at the hospital. This would help re-energize him and reconnect him to the company's mission.

## Why Communicating as a Technology Leader Is More Than Just Sharing Information

Dick Daniels's approach highlights a truth every technology leader should internalize: communication isn't merely about transmitting information; it's about creating alignment, building trust, and inspiring purpose across diverse audiences. Leaders must do more than deliver updates or project plans in IT organizations where technical complexity, business priorities, and human factors intersect. They must intentionally create frameworks that allow their message to resonate with technical specialists, business partners, and senior executives alike.

Through structured leadership team alignment, repetitive reinforcement of strategic priorities, or personal efforts to reconnect with the organization's mission, successful communication transforms a group of specialists into a unified, purpose-driven team. Mastering this skill isn't optional for technology leaders; it enables organizational performance and enduring cultural strength. As Daniels' experience shows, when employees truly understand how their work contributes to something larger, the IT organization executes better. It becomes a place where people want to stay, grow, and be a part of something bigger than themselves.

## CHAPTER 10

# Impacting the Team: What Did We Learn

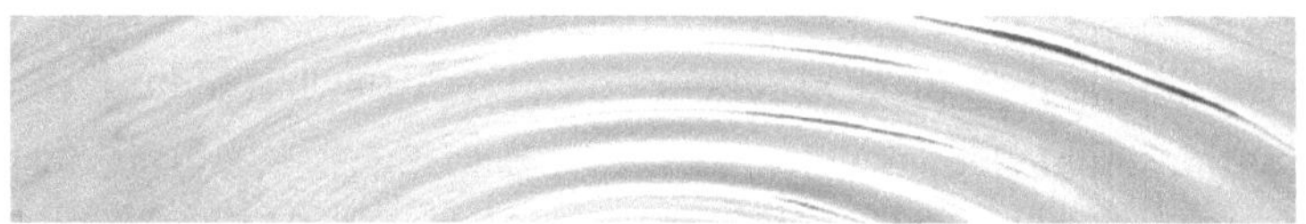

## Motivating the Team Behind the Technology

Grace Hopper's timeless wisdom, "You manage things; you lead people," reminds us that managing systems is critical in high-pressure, fast-moving environments, but building trust, fostering resilience, and cultivating engagement drive long-term success. While tools and processes may keep the lights on, people fuel innovation, adapt to change, and carry organizations forward. Leading them with empathy, clarity, and purpose sets exceptional tech leaders apart.

## The Research Speaks: People-First Leadership Works

The data presented in this section provides compelling evidence for the people-first approach. Leaders from "Best Places to Work in IT" organizations don't just talk about valuing employees; they demonstrate it through their actions. They prioritize supervisor quality (+12%) and competitive compensation (+14%) more than their peers, understanding that a strong foundation of trust and fair treatment creates the conditions for everything else to flourish.

Perhaps most tellingly, these exceptional leaders maintain a balanced approach to Herzberg's hygiene factors and motivators, showing only a 6-point difference in their focus between the two categories, compared to a 16-point gap among their peers. This balance isn't accidental; it reflects an understanding that sustainable workplace excellence requires attention to both the foundational elements that prevent dissatisfaction and the motivational factors that drive engagement and growth.

## Beyond Theory: The Lived Reality of Transformational Leadership

The stories shared in this section, from Steve Zerby's decade-long dedication to creating jobs centered on people at Owens Corning, to Graeme Thompson's mission-driven approach at Informatica, demonstrate that people-first leadership isn't just an idealistic idea. It's a practical approach that delivers measurable results: multiple Best Places to Work awards, talent retention, and organizations that succeed despite periods of disruption and change.

The banking CIO's four-stage transformation journey illustrates how thoughtful investment in the development of people can impact an organization. By cultivating the right mindset, investing in relationships and a strong foundation, and finally focusing on empowerment and capability-building, she didn't just lead a high-performing IT team; she developed a group of change agents ready to challenge the status quo and drive meaningful transformation.

## The Cascading Effect of People Development

These leaders are exceptional because their approach creates cascading effects throughout their organizations. When Hans Keller prioritizes character over competence in hiring, he doesn't just get better team members; he gets leaders who model those same values for others. When Dick Daniels requires his infrastructure teams to walk in their users' shoes, he doesn't just improve service delivery; he creates technology professionals who see themselves as part of a larger mission to improve healthcare.

This explains why 75% of Best Places to Work organizations maintain their excellence across multiple years, and remarkably, 41% do so even when senior leadership changes. People-first leaders don't just build great teams; they build self-sustaining cultures that outlast their tenure.

## The Three Pillars of People-First Technology Leadership

Based on the examples and research presented, three core pillars emerge for technology leaders committed to putting people first:

1. **Listen to Learn, Not to Judge**. The most effective leaders treat every interaction as an opportunity to gain insight; whether Graeme Thompson conducts listening tours to understand Informatica's culture, Hans Keller engages directly with teams to learn about their challenges, or the banking CIO synthesizes feedback from multiple sources to uncover performance gaps. They listen not to confirm their assumptions but to truly understand what motivates, concerns, and inspires their team members and business partners.

2. **Communicate to Connect, Not Just to Inform**.
   Dick Daniels' framework illustrates that effective
   communication in technology organizations involves
   more than just sharing information; it's about building
   emotional and intellectual connections that foster
   understanding and engagement. Whether helping team
   members see how their work contributes to patient care
   or ensuring the entire leadership team is aligned and
   speaking with one voice, the goal is clarity, alignment, and
   a shared sense of purpose.

3. **Act with Courage to Do the Next Right Thing.** People-
   first leadership often requires making difficult decisions
   for the long-term health of the team and success of the
   organization. Take the banking CIO who made the tough
   call to remove 30% of her leadership team. This isn't about
   pleasing everyone; it's about having the conviction to
   make choices that serve the greater good of the team and
   mission of the firm.

## The Fork In the Road: The Choice You Face

As you reflect on your leadership journey ahead, you face a fundamental choice. You can focus on technical excellence, project delivery, and operational efficiency, the "managing things" approach that brought you this far. Or you can embrace the more challenging but ultimately more rewarding path of "leading people."

This choice will define your effectiveness as a leader and your impact on those you lead. Twenty years from now, the specific technologies you implemented will likely be obsolete.

The systems you built will have been replaced. The projects you delivered will be distant memories. But the people you developed, the careers you launched, the confidence you instilled, and the culture you created; those impacts will last for decades.

## Impacting the Team: A Final Reflection

When our son Patrick was contemplating his career, I asked him: "What do you want to be known for in the future?" This question isn't about public recognition; it's about identifying what you're so passionate about that it becomes evident to others. Over years of mentoring, I've learned that everyone's answer is different: some want to be known for their technical expertise, others for developing people, living out their faith, or maintaining work-life balance. Understanding this helps focus your professional and personal development and ensures you work for an employer that supports your passions.

I asked Patrick that when evaluating his current employer or considering new opportunities, first ensure the basics are covered: fair compensation, good relationships with colleagues and supervisors, and a positive working environment. These are the "table stakes" that prevent job dissatisfaction. Then assess whether they enable what you want to be known for: Do you enjoy the work and connect with the organization's mission? Can you develop in areas that matter to you? Are your accomplishments acknowledged meaningfully? Will strong performance lead to new responsibilities and clear paths for career advancement? Your career will never follow a straight path, but maintaining focus on what you want to be known for helps you navigate those inevitable twists and turns.

As leaders, we should understand what each team member wants to be known for. Look for opportunities to align their daily work with their goals and make this an ongoing dialogue that informs performance reviews, project assignments, and development planning. Consider which motivators and hygiene factors are most important to your team and whether these needs are being addressed. Leaders who master this approach retain passionate employees who become organizational advocates, transforming your workplace into a magnet for top performers.

# PART THREE

# IMPACTING THE PROFESSION: EXPANDING & DIVERSIFYING THE TECH WORKFORCE

## CHAPTER 11

# Impacting the Profession: Why It Matters

"Diverse teams make better mousetraps."
**—Scott E. Page**

These words reflect what leaders who focus on innovation and creating great workplaces understand: having a lasting impact depends on access to talent with diverse backgrounds, experiences, and perspectives. However, they acknowledge an ongoing challenge in the technology field of attracting and retaining a deep and diverse talent pool required to sustain their efforts.

Forward-thinking technology professionals see it as their mission to develop the next generation of professionals. They actively work to expand the talent pipeline by engaging students earlier in their education and creating opportunities for individuals from underrepresented communities. Once these individuals join the field, leaders support their ongoing growth and provide the assistance needed to help them stay and flourish in the profession.

## A Trip Back in Time to Fix the Future

Understanding the history of the technology profession is the first step in addressing the gap in talent diversity. Whether you have been in the field for 4 years or 40, you may not know its history and the impact that the rapid evolution of technology has had on the people who have made it their career.

Computing is one of the newest professions, with origins dating back to the 1950s. By contrast, fields such as accounting, procurement, marketing, manufacturing, and even customer service have been established since the 1800s, giving them decades, if not a full century, more time to mature. More importantly, few occupations can claim the level of disruption that the technology field has experienced during its formative years.

These disruptions included dramatically expanding the U.S. IT workforce between 1970 and 2000. According to U.S. Census Bureau data, IT workers increased from about 450,000 in 1970 to 3.4 million by 2000, a growth of more than 650%. Digital technologies spread across industries, fueling this dramatic growth. Yet as the profession grew rapidly, it also gave rise to unforeseen challenges and complexities that continue to shape the field today.

## Growing Pains In the Profession

The field's rapid growth resulted in several issues beginning in the mid-1980s, many of which still have a residual impact. Universities struggled to meet surging demand for those seeking an education in computing. Additionally, there was a brief period of approximately five years from the late 1990s to the early 2000s, when a "perfect storm" of issues arose that would raise concerns about the stability of a tech career. Concurrently, the industry faced the Year 2000 (Y2K) problem, the rise and fall of the dot-com bubble, and a wave of technology jobs moving offshore.

## Struggling to Keep Up with
## the Demand for Talent

While the growth of those entering the IT workforce was impressive, especially between 1980 and 2000, the demand for computing education far exceeded the supply of qualified college-level faculty. **Figure 5-1**, using data from the U.S. Department of Education, illustrates the steady increase in computer and information sciences degrees awarded from the early 1970s to the mid-1980s, when the numbers started to decline.

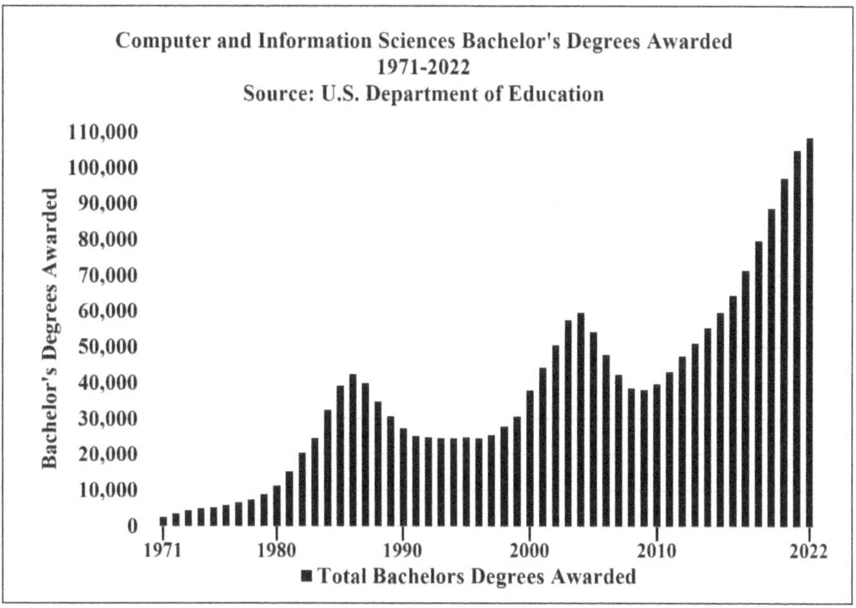


*Figure 5-1*

When computer science programs could no longer accommodate growing student interest, universities began limiting enrollment and implementing strict admission standards. **Figure 5-1** reveals the dramatic impact of these constraints, showing that bachelor's degrees in computer and information sciences declined sharply beginning in 1987. It wasn't until 2001, 15 years later, that graduation rates returned to a similar level as in 1987.

## Y2K

While strict admissions standards limited the supply of new professionals, a global issue arose known as Y2K, short for Year 2000, and started attracting widespread attention in the early 1990s. The Y2K issue arose because programmers commonly used two digits to represent years in computer systems. Dates formatted as mm/dd/yy meant that January 1, 2000, appeared as 01/01/00 and could be interpreted as either the year 2000 or the year 1900. This could have significant implications, especially when using the date for calculation purposes.

The Y2K challenge significantly increased the demand for skilled technology professionals and highlighted the critical need to maintain and modernize legacy systems. Once organizations recognized the global scale of the problem, they faced a surge in the demand for computing resources to repair or replace vulnerable systems. Companies replaced many proprietary applications with customizable ERP systems. At the same time, in-house technology teams had to shift their focus from developing new capabilities to prioritizing the remediation of legacy systems to avoid potential disruptions.

## The Dot-Com Era

As companies invested in Y2K remediation, a surge in internet-based startups sparked a wave of optimism and increased investment in the tech industry. Dot-com companies across various sectors, including online retail and telecommunications, attracted high stock valuations and a strong demand for technology professionals to reach their financial goals.

Demand for software developers, network engineers, and web designers surged, along with salaries and signing bonuses. Young professionals poured into the industry, drawn by stock options and explosive growth potential.

By the early 2000s, the bubble burst. The collapse had immediate and painful consequences for the computer technology profession. Companies conducted mass layoffs across IT departments, hitting dot-com firms especially hard. This period shattered confidence in technology careers, particularly for those who had entered the field during its rapid expansion.

The aftermath extended beyond the immediate job losses and financial instability. For many traditional business leaders, the rise and fall of the dot-coms seemed to confirm their skepticism about the internet's business viability. This led to a pullback in e-commerce investment and a slowdown in digital transformation initiatives within established companies, forcing many IT organizations to reduce their budgets and headcount.

## IT Offshoring

During the rise in demand for technology talent driven by Y2K and dot-com ventures, along with a decline in U.S. university graduates with computing degrees, a new approach to resourcing tech initiatives was emerging, known as offshoring. Countries like India, China, and Eastern Europe became hubs for outsourced technology services, offering scalable workforces at significantly lower costs. A global shortage of IT professionals accelerated this shift. Advances in telecommunications infrastructure and widespread internet connectivity enabled leaders to manage remote teams effectively across time zones.

Once Y2K concerns and dot-com businesses faded from the headlines in the early 2000s, offshoring began accelerating rapidly and attracted more attention. When technology departments faced pressure to cut costs, offshoring work became an increasingly attractive option.

## Impact of the "Perfect Storm"

The rise in offshoring of IT jobs, combined with the dot-com bubble burst, caused major concerns about the long-term prospects of a technology career. As a result, enrollments in computer-related degree programs at U.S. colleges and universities dropped significantly.

In **Figure 5-1**, the impact of this decline is evident with the reduction in the granting of bachelor's degrees in computer science beginning in 2005. It took another eleven years before the rate of computer and information science graduates reached the same level.

A closer look at the numbers reveals an ongoing issue regarding women entering the technology profession. **Figure 5-2** contrasts the percentage of female computing graduates with the total and shows that, after peaking at 37% of all graduates in 1985, the number of women earning bachelor's degrees in computing steadily declined and has not returned to those levels several decades later.

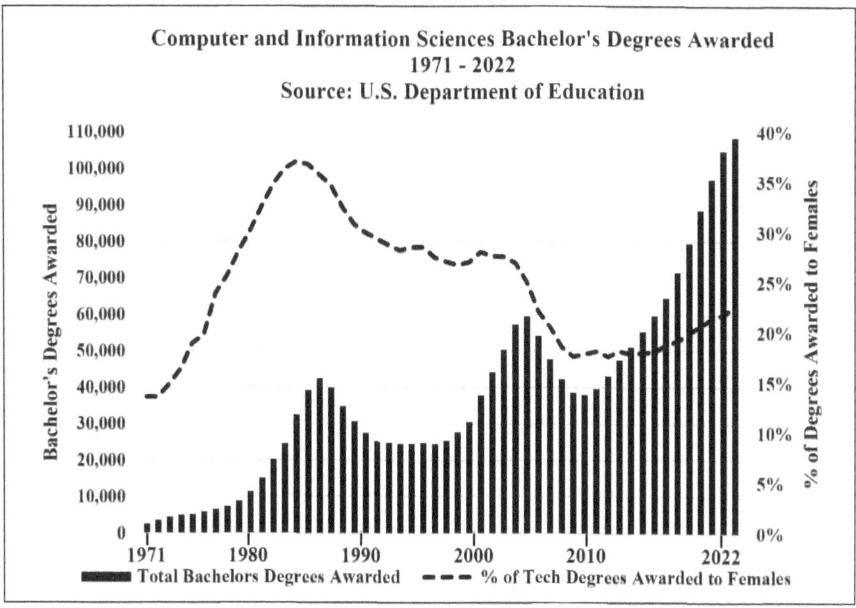


*Figure 5-2*

The first decline in the percentage of females earning computer and information science degrees occurred with the implementation of strict admission criteria in the mid-1980s. The second significant decline occurred following the dot-com bubble burst and offshoring of tech jobs in the early 2000s, reaching a low of 18% in 2008. Although the percentage of women graduating with computing degrees has risen, in 2022 it was still only 22% of all graduates.

## A Deeper Dive: Understanding the State of Representation in 2022

The data from 2022 presents a stark picture and a powerful call to action for today's technology leaders. **Figure 5-3** shows a large gap between those earning computer science and information science (CS/IS) degrees and the broader population.

| Representation by Group in U.S. Population, Bachelor's and Computer Science/Information Science (CS/IS) Bachelor's Degrees (2022) | | | |
|---|---|---|---|
| Group | Population[1] % | All Bachelor's Degrees[2] % | CS/IS Bachelor's Degrees[2] % |
| Male | 49% | 42% | 78% |
| Female | 51% | 58% | 22% |
| White | 59% | 59% | 51% |
| Black | 13% | 10% | 10% |
| Hispanic | 19% | 17% | 13% |
| Asian | 6% | 9% | 21% |

[1]Source: U.S. 2020 Census  [2]Source: National Center for Education Statistics (NCES)

*Figure 5-3*

The data highlights significant disparities in bachelor's degree completion among different gender and racial groups in the CS/IS fields. While women slightly outnumbered men in the general population and earned 58% of all bachelor's degrees, they represented only 22% of CS/IS graduates in 2022. Conversely, men, who made up just under half of the population, earned 42% of all bachelor's degrees but accounted for 78% of the technology degrees awarded. This imbalance highlights a persistent gender gap in technology-related education, suggesting barriers and cultural dynamics that continue to discourage female participation in technical disciplines despite their overall academic success.

Racial differences are similarly striking. White students, who represent about 59% of the population, account for 59% of all bachelor's degrees and 51% of CS/IS degrees. Black and Hispanic students, who together comprise roughly 32% of the population, are notably underrepresented in these technology degrees, earning only 10% and 13% respectively, a gap even more pronounced than their underrepresentation in bachelor's degrees overall. A notable statistic is that although Asian Americans make up 6% of the population, they earn 9% of all bachelor's degrees and 21% of all CS/IS degrees.

While these statistics reflect the situation in the United States, gender and racial gaps in technology education are also found in other regions of the world. The specifics regarding gender and racial underrepresentation will vary by country; however, technology leaders should view these numbers as more than just statistics. They represent a talent pipeline that will limit the profession's future. Leaders in the field have a unique opportunity and responsibility to change this. By investing in outreach, mentorship, and inclusive hiring, leaders can help ensure that tomorrow's technology workforce represents the diversity of the world it aims to serve.

## Where to Begin Making an Impact

If you're a technology leader who wants to shape the profession's future, a natural question arises: *Where should you begin?*

**To explore this, a survey was sent to tech executives asking a simple question:** "Who should technology leaders focus their efforts on to help build a deeper and more diverse talent pool to support the future of the computing profession?" Respondents could select any number of these five options that they deemed important.

1. Primary and secondary school students (ages 6 to 18)
2. College and university students
3. Experienced professionals (both in and outside of tech)
4. Other (with the opportunity to specify)
5. None of the above – this should not be a priority for technology leaders

After receiving responses from more than 660 leaders from 48 countries, the results clearly show:

- 75% of respondents identified school-age and college students (options 1 and 2) as top priorities
- 60% suggested focusing on experienced professionals
- 20% selected "Other"
- Only 1% thought this effort should not be a focus for tech leaders

Notably, 79% of participants selected more than one option to focus on, averaging 2.3 per respondent. This shows that most technology leaders see the need for a broad, multi-pronged approach to nurturing the next generation of tech talent. Appendix C provides further details on the survey, including a breakdown by title, gender, industry, and company size.

While the data exposes significant challenges, it also points toward meaningful opportunities for progress. The past disruptions that led to current barriers also show the profession's remarkable ability to evolve when thoughtful leaders step forward. The statistics on underrepresentation reflect more than trends; they highlight tremendous untapped potential waiting to be developed. Fortunately, professionals worldwide are already working to change these patterns. Through mentorship, outreach, and creating new pathways into the field, they're building a diverse talent pipeline to strengthen the profession. Their experiences demonstrate how individual dedication can create lasting change and inspire others.

## CHAPTER 12

# Impacting the Profession: How Do They Do It?

Spanning efforts that support grade school and university students, working professionals, and adults pursuing a new career in technology, the following stories demonstrate the many ways leaders are making a difference. Whether through large-scale initiatives or personal mentorship, each example inspires those seeking to give back in meaningful and lasting ways.

More details about the five organizations highlighted in this chapter are available in Appendix D.

# DR. PHILILE MKHIZE
## Lifting as She Leads: A Mission to Empower Girls and Women in Tech

> "As long as it involves technology, people, and
> their advancement, I am there."
> **—Dr. Philile Mkhize**

Growing up in the small rural village of Mkhazane in the Kwa-Zulu-Natal province of South Africa, Philile Mkhize never could have imagined the successful career she would achieve in the technology field. The approximately 500 km (300 miles) distance between her village and Johannesburg represented more than just geography; it was a gulf between two worlds. As a young girl, it would have been unimaginable to think she would one day work in the banking, insurance, and telecommunications industries, or co-found a FinTech startup that helped those without bank accounts establish credit. Growing up, her vision of the future was shaped by what she observed the women doing in this humble and poor community, with little awareness of the opportunities beyond her village borders.

The inspiration she gained from one of those women, her grandmother, motivated her to pursue her dreams and break the cycle of poverty that had afflicted so many members of her village. Mkhize's grandmother would tell her that "her brain was just as good, if not better than anyone else in the room, and that her perspectives and opinions matter." She would remember her grandmother's words if she ever doubted herself and wondered if she belonged somewhere.

Mkhize's grandmother emphasized the importance of education and did everything she could to ensure she received the proper schooling. She left home every morning and walked 20 miles (12 km) to attend school. Walking this distance to and from school each day helped to reinforce how important it was to her family for her to be successful. After completing high school, she realized that walking that distance daily to attend class was the first step toward a career that could extend well beyond her village. She knew she could "shift her boundaries and thinking" to go much further than what was common in her village.

## Finding the Way to a Technology Career

Although she had no specific career plan, Mkhize left her village and enrolled at the University of Johannesburg. What she did know was that she wanted an education that would enable her to have a career that would help to change the lives of her family. She had an aptitude and passion for math and science, prompting her to pursue a degree in mining. Historically, South Africa's mining industry restricted women from working underground; however, this changed with the implementation of the Mines Health and Safety Act of 1996, the same year Mkhize began her university studies.

Halfway through one of her first courses, the class was invited to visit the operations of one of the largest mining companies in South Africa. Mkhize didn't fully comprehend what it meant to work inside a mine, and this trip was instrumental in helping her realize that this was not the environment she wanted to work in for her career. As she emerged from the underground, she realized she needed to explore other career options.

Mkhize spent several weeks considering other professions, focusing on which one could help her achieve her goal of impacting her family. At this pivotal moment, she received career guidance from the university and decided to take a leap of faith by pursuing a degree in technology. She had a strong interest in technology's impact on her and her country's future. She was also interested in combining her interest in tech with her interest in business. Consequently, she decided to make the "life-changing decision" to pursue a degree in Informatics, which involves the study of computer science and its application to solve real-world problems.

## Playing Catch-Up

Despite her interest in the subject, Mkhize had never been exposed to computers. She had no experience typing on a keyboard, let alone writing code. She recognized that this would require additional effort since many of the other students had been exposed to computers in high school. She believed that she needed to work harder than anyone else because this was the path she had to take to succeed. As a result of her efforts and the support she received, she was one of the first black females to graduate with a degree in Informatics at the university.

When she graduated, Mkhize had three job offers, and the one she chose would play a crucial role in her career development. During her first six months at the job, she saw how technology could impact a business and how computers could facilitate payment and banking transactions.

This was an exciting time in Africa to be in the tech field. Telecommunication companies, including Vodafone, expanded rapidly across the continent, enabling new economic growth. Mobile phones enabled people in rural areas across the continent to access financial tools, making remote banking transactions, including savings, loans, and electronic payments, more convenient. Technology also enabled more women in these communities to participate in the broader economy by removing barriers to education and financial independence. The potential for technology to alleviate poverty in many small villages across the continent, including where she grew up, ignited Mkhize's interest and continues to fuel her passion for her chosen profession.

## Support from Those Who Saw Her Potential

From her grandmother to teachers and professors at school, Mkhize received support from people who recognized her potential despite her humble beginnings. They believed in what she could achieve, and their encouragement increased her confidence and academic success.

That support continued as she began her professional career. She has had leaders who "took a bet on her" and supported her throughout her career. Whether it was a line manager or members of her work network, she had mentors who invested their time to help her succeed. As a result, Mkhize progressed through a wide range of technology roles, including project manager, systems architect, business intelligence manager, and several executive-level positions.

One of her earliest advisors was Christi Maherry, a highly recognized leader in the cybersecurity industry. Maherry believed in Mkhize and her dream, and that dreams like hers mattered. Twenty years later, their relationship continues. The only difference is that many of their discussions now offer Mkhize an opportunity to serve as a reverse mentor and help Maherry with some of the challenges she faces in her work.

Mkhize feels a deep sense of responsibility to give back to the profession by helping aspiring technology professionals, thanks to the support she received from Maherry and others during her academic and professional journey. Inspired by her mentors, she is committed to encouraging students to enter the field and sharing her experiences with those already in the profession to help them navigate their journeys and avoid some of the challenges she faced. Her commitment to supporting aspiring technology professionals has allowed her to make a meaningful impact on women of all ages worldwide.

## Seizing the Opportunity to Pay It Forward

In 2017, Mkhize was selected to participate in a global women's conference in Silicon Valley, located in the South San Francisco Bay Area of California, which is recognized as the hub of the tech industry. When she first arrived, she was amazed to find robots performing everyday room service tasks and using self-service registers at the retail stores she visited. This was the work often performed by many young girls in South Africa who could not afford a university education. If these became commonplace in her country, what work would remain for these women, and how would they survive?

At that moment, she realized that she wanted to make a difference in the lives of these young women. Rather than passively watching technology replace their work as housekeepers and cashiers, Mkhize wanted to inspire them to pursue a career in technology, enabling them to acquire the skills necessary to develop code for those robots and cash registers.

While on the same trip to Silicon Valley, she attended an event held at Google headquarters that would provide the solution she was looking for. The Technovation Challenge is a global competition encouraging young women to use technology to solve real-world problems. The competition involved more than 11,000 girls from 103 countries, who formed teams to tackle issues related to peace, poverty, the environment, equality, education, and health. The event, hosted by Google, was the final round of the competition, during which finalists presented their ideas to tech leaders and professionals.

The apps competing for the top prize aimed to address issues such as human trafficking in Kazakhstan, sign language education in Armenia, and access to healthcare information for pregnant women in rural India. Mkhize was immediately inspired by what she saw and wondered how she could support Technovation in South Africa. She wanted to help young girls learn how to use technology to solve problems in their communities, scale their innovations, and commercialize them.

**Getting Involved to Support Girls in South Africa**

Mkhize began her involvement with Technovation by supporting a group of girls in Soweto, a township in Johannesburg. These young women in grades seven and eight wanted to develop an app that targeted gender-based violence. This was a growing concern throughout the country, as highlighted by domestic violence statistics documented in the 2016 South Africa Demographic and Health Survey. The report noted that one in five women (21%) had experienced physical violence in their lifetime by a partner, with 37% of those cases resulting in injury. In 2017, this issue in South Africa drew global attention after reports highlighted the significant increase in violent crimes committed against women by their partners.

One of the girls on the team had a mother who was a victim of this violence. She noted that when her family visited the police station to report the crime, the account of what had occurred changed because the father, who committed the crime, was present. Through her involvement with the Technovation program, this young woman recognized the potential of technology to help address this issue.

Mkhize mentored and advocated for the team, motivating them and providing the necessary resources to complete their project. Her guidance enabled the team to design and develop a mobile application that alerts the police when a victim of violence shakes their phone. If the woman experiencing this crime dropped her phone on the floor, the app would notify the police of her location and inform them that the situation was escalating, thereby raising the priority for the police response. The team's innovative solution garnered the interest of others, including that of South Africa's Department of Women and Children, as a potential solution to this growing problem.

Mkhize is proud of the foundation that the Technovation Challenge has established for these young women. It has helped them understand the power of technology and, more importantly, how it can address the problems they face daily. She is inspired by observing how these young women think and collaborate to address significant real-world problems, especially in contrast to the minor issues many of us face in our work.

She has since increased her involvement with Technovation and is the head judge for entries submitted worldwide. Her role is to review the entries and assess the problems they address and their proposed solutions. She also evaluates how the teams of 5-7 girls have thought through their solutions and how they have matured as a team over the 3-5 month period they work together. The experience for these young women is more than just learning about technology; it also involves working with and collaborating with others. Mkhize notes that once she started judging the projects and the problems the young girls were working to address, "you cannot stop; you want to see what else they are capable of achieving."

The impact of Technovation extends well beyond the competition itself. For many of these young women, it becomes a launching pad for future careers in technology and science. As of

2024, Technovation alumnae number more than 55,000 women and girls over 18. An impressive 87% of the young girls who have participated in the program plan to or are pursuing STEM degrees, with 54% of those focused on tech fields such as computer science, artificial intelligence, and robotics.

## Supporting Women throughout Their Careers

In addition to her work with Technovation, Mkhize is involved with Women in Technology (WIT) in South Africa. WIT's mission is to reduce the gender gap in the STEM field by encouraging more women to pursue careers in this field. Her involvement with the organization focuses on ensuring that once women decide to pursue STEM careers, they have the support to stay in the profession. She wants to serve as a role model for young women, demonstrating what can be accomplished with a career in technology.

She does that by mentoring women in university and those already working in the field. In four years, she has provided career guidance to eighty women. Mkhize follows a principle of "One hour, one mentee, one week, the power of three." The idea is to amplify her impact. If she impacts one person in one hour a week, she can double the effect by spending two hours with two mentees in two weeks.

Mkhize is passionate about impacting the number of women in the field. She notes, "We all have experiences we can share with others, and it doesn't need to be through a formal mentoring relationship." It can be as simple as answering a young girl's question about her career or an opportunity to help someone think differently about their future, providing the encouragement they need to continue pursuing their dream.

## Committing Time to Make a Difference

Like many industry professionals, Mkhize has a busy schedule of activities both at work and at home. As the Chief Operating Officer for Technology at the pan-African bank Absa Group, she had a full agenda with a role that included leading the technology organization, driving operational efficiency, and providing oversight of Absa's global technology hub in the Czech Republic capital, Prague. Despite her demanding schedule, she still finds time to impact the lives of others who want to pursue and succeed in a career in technology.

Mkhize strongly encourages other technology leaders to focus on the profession's future. She notes, "While you may not see the results of your efforts in your lifetime, your contribution will make a huge difference in future generations. The work starts today."

She believes that impacting one person's life is the most important work that can be done. It's her purpose and what drives her. She strives to leave a lasting impact with each interaction she has with others. Regardless of her other commitments, she will make time to have an impact because she wants another young girl from her village who hears her story to say, "If Philile can do it and I have the same brain, and maybe better than hers, why not?"

The words of encouragement and support that her grandmother gave her will resound for future generations of women in South Africa. She knows that there is no greater impact that she can have than helping others realize their dreams. Like the ripple that a single drop of water creates in a lake, Philile Mkhize will continue to touch the lives of other women, and they, in turn, will do the same.

# CINDY MCKENZIE

## Inspiring University Students to Pursue Careers in Tech

"Besides having my children, this is probably
what I am most proud of."
**—Cindy McKenzie, Co-Founder STEM Advantage**

After decades of being among the few women in corporate boardrooms and CIO meetings, Cindy McKenzie thought she had grown accustomed to being outnumbered. Seeing a cohort of incoming computer science students with only a dozen female faces among hundreds of students ignited a different response: not resignation, but a call for change. That moment of clarity didn't happen overnight. McKenzie's path to becoming an advocate for STEM diversity began decades earlier.

McKenzie studied Economics at California State University, Long Beach, where she also enjoyed taking BASIC and COBOL classes. While working part-time as a bank teller during her senior year, she had her first exposure to the technology field. The bank had just been acquired, and most of its IT department staff had left. She was asked to help convert the bank's systems to those of the acquiring bank. This six-month assignment offered her a significantly higher salary than her position as a bank teller, so she didn't hesitate to accept the opportunity. Upon graduation, the bank was so impressed with her work that she was offered a position as a business analyst in the IT organization.

McKenzie enjoyed working with the business to help them understand their challenges and how IT could solve their problems. She advanced from business analyst to project manager and believes that this experience helped prepare her for future leadership roles by requiring her to manage projects and teams, even though they did not report directly to her.

After eight years with the bank, she rose quickly to vice president of branch automation and business application software. When another bank acquisition resulted in her IT department moving out of Southern California, she pivoted into the entertainment industry. After several different roles across the industry, she eventually became the global CIO of Deluxe Entertainment Services Group, where she was responsible for all of the technology-related functions of the business, including application development, security and compliance, business intelligence, and infrastructure. But even as she reached the executive level, McKenzie couldn't ignore a troubling pattern that had followed her throughout her career: she was consistently one of the few women in the room.

## Career Development with the Help of Mentors

Throughout her professional journey, McKenzie had sponsors and mentors, most of whom were men due to the limited number of women in the field. They believed in her, offered advice, and helped her grow as a leader. While serving as a vice president in the entertainment industry, she worked for a female CIO who served as a role model and provided invaluable insights that guided her.

McKenzie valued her mentors and believed you should be open to learning and getting advice from others, even if they are not in your field. If she saw someone in her company who had an interesting way of dealing with a problem, she would seek their advice on the issues she was facing. She firmly believes in the importance of having and being a mentor. This would ultimately become a core element of the non-profit she would help establish.

## The Call to Action to Create STEM Advantage

McKenzie was well aware that IT leadership remained a male-dominated space, something she was reminded of every time she stepped into a room of tech executives and saw only a handful of women, often just herself. However, the issue became deeply personal during a Parents' Night at the University of California, Los Angeles (UCLA), where her son studied computer science. As she scanned the crowd of hundreds, she was struck by how few women were present. She could count them on two hands. That moment crystallized a truth she could no longer ignore: not enough women were entering the field, and even fewer were staying. She knew then that she had to act to help open doors and build pathways for the next generation of women in technology.

While serving as the Senior Vice President at Fox Entertainment Group, McKenzie attended a Society for Information Management (SIM) meeting and was introduced to Lee Ann Kline. Kline was a technology leader who wanted to start a non-profit to help women and underrepresented minority college students pursue careers in Science, Technology, Engineering, and Math (STEM). She knew that McKenzie could assist her by engaging her connections with other regional industry professionals to support the program. Together with other business, technology, and talent leaders, they co-founded STEM Advantage in 2012.

STEM Advantage focuses on individuals pursuing degrees at the eight California State University (CSU) campuses in Southern California, the largest and most ethnically diverse four-year public universities in the United States. The organization provides a diverse talent pipeline through partnerships with businesses, enabling participants to gain exposure to working in a corporate environment and establish relationships with others in those companies. Most of the students in the program are the first in their families to attend college and don't have access to a network of professionals working in their field of study. STEM Advantage helps them build skills and a network that enables them to obtain internships, which can then lead to full-time employment.

McKenzie notes that, through STEM Advantage offerings, "We can change the trajectory of the students' lives." Many of the program's alumni earn at least $75,000 annually; some are offered salaries exceeding $100,000 for their first job after graduating. As a result, the organization is changing not only students' lives but also the lives of their families.

### Getting STEM Advantage Launched

For the first ten years, STEM Advantage was an all-volunteer organization. During that time, McKenzie and Kline were employed full-time and used any spare time to expand the program. Working evenings and weekends, they developed what would become their flagship offering. The Scholars Program, designed for students in their sophomore through senior years, became the foundation of everything STEM Advantage would build. The requirements to be accepted are that participants must be U.S. citizens or legal residents, major in a STEM-related discipline, and have a 3.0 grade point average (G.P.A.). Once they are accepted, they have access to the following offerings:

- **Internships in technology positions.** Participants are paired with local employers for internships, which may be extended into part-time positions during the academic year. They receive the standard hourly wage for interns. They may rotate through different roles during the internship to gain experience in various opportunities within their chosen field.

- **Mentorships with working professionals.** Students are matched with professionals who work in their chosen STEM fields and provide critical support to help them stay in school and complete their degrees. Mentors offer professional insights, review resumes and LinkedIn profiles, and sharpen interview skills through mock interviews. They also serve as trusted sounding boards for questions about school, work, and life. Mentors and mentees meet for an hour each month, in person or virtually, for a year, with many relationships continuing well beyond graduation.

- **Scholarships.** Each participant in the Scholars Program receives a renewable scholarship to defray the cost of tuition. Along with their internship compensation, the scholarship helps them complete their studies with less debt.

- **Professional and Career Development.** Students learn to communicate confidently and professionally, develop strategies for building their professional networks, and acquire skills for engaging in technical and behavioral interviews. Additional sessions addressing imposter syndrome, financial literacy, and professional etiquette are also offered.

  Participants also attend industry panel discussions addressing topics ranging from women in technology to a "day in the life" in various disciplines, including cybersecurity, data analytics, and software engineering. In addition, panel discussions focus on specific industries in California where students would likely seek employment, including entertainment, aerospace, and healthcare.

McKenzie conducts a mandatory "IT Jobs" session where she guides students through a typical IT organizational chart, highlighting entry-level roles across job families. She broadens their view beyond software engineering, introducing roles in project management, ERP, CRM, and more while emphasizing the technical and interpersonal skills needed to succeed.

In 2022, McKenzie and Kline broadened their reach by launching the Freshmen STEM Career Pathways, focusing on first-year students at CSU locations. It shares elements similar to the Scholars Program, such as career development panels, resume writing assistance, and mentoring circles instead of one-on-one mentoring. The mentoring circles are often led by professionals who are alumni of STEM Advantage and provide these participants with the insight of someone much closer to their age and who has already benefited from STEM Advantage. At the end of their first year, participants with a 3.0 GPA or higher will automatically qualify for the Scholars Program the following year.

Another initiative that feeds the Scholars Program focuses on college transfers with a two-year associate's degree. This initiative educates community college students about professional opportunities in STEM and prepares them to transfer to a 4-year college and complete a transfer to a CSU location where STEM Advantage operates; they are invited to join the Scholars Program. Like the Freshmen offering, the Career Pathways initiative provides members with professional development and mentoring circles. Another focus is supporting these students transitioning from community college to a CSU institution. STEM Advantage works with each person to help identify the four-year university that best meets their STEM interests.

One of STEM Advantage's most impactful offerings is its network of peers and professionals to help participants navigate their collegiate and professional careers. The organization wants students to feel supported and motivated, and it accomplishes this by connecting them with alumni, industry professionals, and hiring managers. Since most individuals involved in STEM Advantage are from low-income homes and are the first in their families to attend college, many lack a network to assist them in their professional journey. This is where STEM Advantage excels by sponsoring networking events and opportunities for alumni to give back by serving as mentors, role models, and advocates.

### Making an Impact

Research that emerged in the year McKenzie and Kline launched their initiative validated their approach in founding their non-profit. In 2012, the United States President's Council of Advisors on Science and Technology published a report titled "Engage to Excel: Producing One Million Additional College Graduates with Degrees in Science, Technology, Engineering, and Mathematics." This report was developed with input from postsecondary STEM teachers, curriculum development experts, higher education administration, and industry professionals. It was driven by the need to address a forecasted demand of producing 1 million additional college graduates in STEM fields over the next decade.

The report states, "Fewer than 40% of students who enter college intending to major in a STEM field complete a STEM degree." It further notes that women and members of minority groups constitute approximately 70% of college students. However, only about 45 percent earn STEM undergraduate degrees, and this figure needs to increase if the demand for STEM workers is to be satisfied.

STEM Advantage addresses these issues by supporting women and the underserved in pursuing and completing their education in a STEM field. In the first 12 years since its inception, more than 1,000 students have benefited from STEM Advantage's offerings. Thanks to a comprehensive strategy that includes mentorship, internships, and scholarships, 100% of these participants have continued their education and graduated with technical degrees, exceeding the national statistics outlined in the aforementioned government report.

STEM Advantage's success is partly driven by over $2.8 million in scholarships awarded to participants. About 300 students are actively engaged in the Scholars Program, while another 200 are involved in the first-year and community college offerings. The organization's success has significantly boosted the interest of prospective candidates, and continued growth in the program will only be limited by the financial and human resources required to support this initiative. For McKenzie, those constraints are simply challenges to overcome, not reasons to slow down.

## Staying Engaged to Continue Doing the Next Right Thing

What drives someone to dedicate their post-corporate years to an all-consuming volunteer effort? For McKenzie, the answer lies in moments that statistics can't capture. She sees her work with STEM Advantage as a "labor of love" that helps her remain energized and engaged in the organization's work. She states, "Besides having my children, this is the thing I am most proud of." She further notes that she is not "nearly as proud of having been one of the few women CIOs as she is of the achievements of STEM Advantage."

Her passion is centered on the students. She states, "When you meet these students and see their lives change, it is inspiring." She shared the story of when she entered an elevator at work and noticed a former STEM Advantage participant and her adult daughter. After they were introduced, the daughter told McKenzie that she wanted to thank her for all that STEM Advantage did for their family. Her mother was planning to drop out during her final year because she couldn't afford the tuition for both of them to attend college. They could complete their education only because of the scholarship money that her mother received from STEM Advantage.

Another inspirational story was shared by a STEM Advantage graduate when he was asked to speak at their annual fundraising gala. The young man was among the most sought-after students, and unbeknownst to many in attendance, he was in a foster care group home when he enrolled in STEM Advantage. His internship enabled him to earn enough money to move from his foster group home into an apartment and live independently.

McKenzie encourages other technology professionals to get involved in similar initiatives, starting small, finding what fits into their lives, and scaling up from there. "When you become a mentor, you build a real emotional connection with students," she says. "You have the power to change their lives." For many, a mentor may be the first person in their lives with experience in the tech field who can offer meaningful guidance and support.

When asked why she isn't spending retirement focusing on traveling or consulting, McKenzie's answer is simple: nothing is more important than helping these students. For her, the greatest reward comes not from accolades or financial gain, but from knowing she's opened doors and changed futures. Her purpose is clear; it's all about the students and making a lasting difference in their lives.

# SIMON REITER
## Mentoring to Develop the Next Generation of Tech Leaders

"As a leader, you are in a privileged position, so it's
essential to remain humble and support the next
generation of leaders in our profession."
**—Simon Reiter**

Within a year, Simon Reiter watched his mentees transform: one overcame her fear of public speaking to address hundreds at an industry conference, another improved her team's reporting to drive company-wide engagement, and a third successfully navigated a career transition to a C-suite role. These outcomes stem from Reiter's dedicated approach to developing future technology leaders.

For Reiter, moments like these validated his core belief: "As a leader, you are in a privileged position, so it's essential to remain humble and support the next generation of leaders in our profession." However, his journey to becoming a mentor and inclusive leader began decades earlier with a Commodore 64 home computer given to him by his uncle.

Reiter grew up in the 1980s in McLaren Vale, the birthplace of South Australia's wine industry. His earliest memories of what he wanted to be when he grew up were a truck driver or a train engineer. As he grew older, he shifted his sights to flying airplanes. However, standing at six feet six inches, he realized his height would prevent him from becoming an airline pilot.

He discovered his passion for technology when his uncle gave him a Commodore 64 computer. He recalls that the system would take 20-30 minutes to boot up after connecting it to the family's television. At that point, he would use the computer until his father wanted to watch the news, requiring him to disconnect the system and patiently wait for another opportunity to access the TV. Despite the challenges of using early home computers, Reiter's enthusiasm for technology remained undiminished. More importantly, these early experiences with patient problem-solving and working within constraints would later shape his approach to mentoring, recognizing that growth occurs through persistence.

## Early Influences

When he entered high school, Reiter had access to many more computers as his interest in technology blossomed. The school's IT manager noticed his interest and asked him to help during lunch breaks. He learned a great deal from this experience, and when he graduated from high school, he was offered a part-time job to support the school's computers. He observes that this experience exposed him to the full spectrum of technology roles, including database administrator, network engineer, and website administrator.

Reiter's first real IT job was with a small school near the Fleurieu Peninsula south of Adelaide, where much of his learning came through on-the-job training. While working at the school, he noticed many students were interested in technology. Like the IT manager at his high school, he helped foster that interest by assigning them small tasks. This experience helped him realize that nurturing the technical skills of these students could bring him greater fulfillment than merely mastering the technology and would ultimately influence his decision on where he wanted his career in the profession to go.

He noticed a pattern as he progressed through different technical roles. Whether helping students at the school near Fleurieu Peninsula or solving complex technical challenges for colleagues, his greatest satisfaction came not from mastering the technology itself, but from watching others grow and succeed. This realization solidified when he faced a career-defining choice: to continue developing his technical expertise or to move into technology leadership. The decision, when it came, felt inevitable.

## Becoming a Better Leader by Being Open to Feedback

Reiter understands that his career growth came from three sources. First, the best managers had articulated how he could improve his performance by providing clear feedback. They didn't shy away from having hard conversations about where he could be more effective in his role. He also had industry mentors who took the time to understand what he wanted to achieve in his career and offered the guidance he needed to reach those goals. As he advanced to higher levels, he sought professional coaches who challenged him to raise the bar on his personal and professional goals.

Reiter understands that an external perspective is essential for developing as a technology leader. Without it, he notes, leaders can become too immersed in their ideas to reflect on them effectively. Having someone to talk to and challenge your ideas will force you to pause and reflect on them more thoroughly. As he used this feedback to grow his career, he knew he wanted to offer the same guidance to others seeking career advice.

## Becoming a Better Leader by Being A Mentor

Having experienced firsthand how feedback, mentorship, and coaching had accelerated his career growth, Reiter was determined to pay it forward. When he moved to Melbourne, Australia, and became the CIO of a financial services firm, a colleague introduced him to Women 4 STEM. Established in 2005, this not-for-profit takes a unique approach to encouraging and supporting women of all ages to enter and remain in STEM careers.

Women 4 STEM offers programs that inspire schoolgirls to consider careers in STEM, provide transitional support for university students entering the workforce, and maintain ongoing engagement with women in roles from entry level to senior leadership to encourage their retention in the profession. After learning about the holistic approach that the organization took to support women in pursuing careers in tech, Reiter knew he wanted to get involved.

He recognizes that the profession's lack of female diversity is partly due to the negative perception of technology professionals, as characterized by movies like "Revenge of the Nerds" or television series like "The IT Crowd." Other factors can lead women to pursue a different profession, including the lack of early exposure to what a career in technology can offer and the lack of visible female role models.

Reiter also appreciates that the more diverse your teams are, the better you are at finding solutions that yield the best outcomes for your firm. He observes, "When you are challenging, improving, or transforming in your role, you are more successful with greater diversity on your team." Team diversity involves members with different backgrounds, including experiences, education, geography, and gender. Just like having a mentor who asks thought-provoking questions, having a diverse team forces a leader to consider different viewpoints that may challenge their conventional thinking.

Reiter is passionate about improving opportunities for women in computing so that they can experience the same enriching career that he has. As a father of two daughters, he realizes that if current trends continue, their STEM career opportunities might be limited, and if he isn't part of the solution, he is part of the problem. When his daughters complete their secondary education and consider their career options, he would like them to have a wide range of possibilities, including a career in technology.

Understanding the rationale for diversity was one thing; actively creating change was another. For Reiter, the path forward was clear: he needed to move beyond advocacy to action. His approach to enacting change would be through mentoring via Women 4 STEM's mentor(SHE:) program. Through this program, he dedicates his time to mentoring women who are senior managers aiming for higher career positions, including C-suite roles. His conversations cover topics such as how to ask for a pay increase and navigate job changes. He continually learns from his mentees and often assists them with challenges he has never encountered. These issues can prompt him to reflect on his leadership style and approach with his team members in similar situations.

While the mentoring relationships last for one year, Reiter notes that you become deeply invested in your mentees and will want to track their progress long after that. He stays in touch with many of those he mentors and enjoys learning what they have accomplished personally and professionally. Hearing the results of his mentoring helps him validate that his advice is valuable and practical.

One example is a mentee with a management role who feared public speaking. Now that she was in a leadership position, she had to be able to voice her opinion in meetings and speak in front of an audience. By the end of the mentoring year with Reiter, she was asked to be a member of a panel discussion at an industry conference, where she spoke in front of hundreds of attendees. His mentee overcame her apprehension about voicing her views in public with his assistance. As a result, she received recognition within her company for her participation and an email congratulating her from the firm's CEO. Afterward, she expressed her gratitude for his assistance, remarking that his support in conquering her fear of public speaking had helped her accomplish several goals, including landing a project she wanted to secure.

## Balancing Personal and Professional Demands while Making an Impact

Like most technology leaders, Reiter's job is demanding. Fitting a mentoring session into his schedule can be challenging due to his work, travel, and family commitments. He balances running his company's technology operations with mentoring responsibilities, viewing both as essential. As a technology leader, he knows that part of his responsibility is to help the next generation and build brand awareness for the profession. He understands that this is just a part of being a C-level executive and that he must prioritize mentoring while finding time to incorporate it into his schedule.

## Mentoring Pays Dividends Now and into The Future

While leading transformation projects, replacing core systems, and achieving savings are part of every technology leader's core responsibility, mentoring can be among the most rewarding.

Reiter notes, "By the time my children enter the workforce, the people I am mentoring now will be the good leaders of the future who will give my children the same opportunities to grow their careers as I was given. It is a self-fulfilling circle."

One advantage of mentoring is the opportunity to enhance your leadership skills by gaining insights into your organization from your mentee that your team might not share. For example, Reiter had a mentee who managed a service desk with 60 agents. She was concerned that no one in her company read the lengthy monthly report her team produced with service desk statistics. He challenged her to think about why anyone would care to read it. He asked, "Why would the CIO care if your team closed 20,000 tickets in the past month? What does the report and the data contained within it mean?"

The mentee considered that feedback and revised the report from an 80-page document over the next several months to a simpler dashboard linked to business outcomes and KPIs. Her users appreciated the updated report format and started interacting with her team more as they began to understand what they could do. At the same time, he started to consider how this could apply to the service desk in his organization. It gave him insight into the issues his service desk manager might grapple with that he would never hear about.

Another insight Reiter has gained from mentoring is that he has learned how women can approach new job opportunities differently than their male counterparts. Many of his mentees have looked at the requirements for a new job opportunity and felt they needed to fulfill all the listed requirements before applying. Their male counterparts may only meet a few requirements and decide to apply for the position anyway. He has helped his mentees understand that to obtain the job they want, they need to believe in their abilities to meet all or most of the job requirements, even if they haven't already performed that job.

Understanding how many women view their abilities based on written job requirements has helped Reiter in his interviewing and hiring process. It has helped him rethink how a job listing should be written if he wants to attract the broadest possible set of talent. As a result, he has transformed his leadership team from being exclusively male to 50% female. He has also learned to ensure that there is always at least one woman on an interview panel for jobs in his organization to help provide a broader perspective on the candidates and their abilities.

Simon Reiter's career aspirations evolved from steering trucks, trains, and planes to steering careers in technology. He has helped shape others' futures through mentoring and has transformed himself into a more inclusive and effective leader.

# JULIE RAGLAND
## Opening Doors: Expanding Access to Tech Careers

> "You get exposure to people who are changing their lives, their communities, and contributing to companies in ways that refresh and challenge our thinking."
> **—Julie Ragland**

What if the solution to the tech industry's talent shortage isn't in coding bootcamps or computer science programs, but in the untapped potential of adults who have already shown resilience and the ability to overcome adversity? For Julie Ragland, this question became a guiding insight. As a member of the Advisory Board for i.c.stars, she saw firsthand the determination and grit of individuals often overlooked by traditional hiring pipelines, yet full of the talent the industry desperately needs.

Founded in Chicago in 1998 by Sandee Kastrul and Leslie Beller, i.c.stars was created to open doors into the technology field for underserved adults, providing a transformative path into IT careers. After two decades of impact and hundreds of program graduates, the organization expanded to Milwaukee in 2018, just as Ragland stepped into her new role as Chief Information Officer at International, the truck and bus manufacturing company. She was then invited to help guide the nonprofit's mission in its new city, a role that aligned perfectly with her vision for inclusive talent development in tech.

Ragland recognized that the demand for technology talent was an ongoing challenge for IT leaders like her. To address that shortage, she saw i.c.stars' mission as a way to identify and nurture talent from non-traditional and underrepresented groups.

To appreciate her interest in serving on the Advisory Board for i.c.stars, it is essential to understand her unconventional path to becoming a CIO. Despite her father being a computer science professor, her initial career pursuit was not in computing. She wanted to pursue a business career and earned her undergraduate degree in accounting. Her first job was at a software company that sold accounting software, where she could combine her education with technology.

Ragland reflects on this experience when she encourages people to consider a career in technology. She asks them to think about technology as the "and" part of what they do, much like when she combined accounting and technology in her first job. She tells people they can combine their passion for clothing design and technology, or animal research and technology. It is an extra layer that can be added to your career interest to give you a different way to see your profession and to differentiate yourself from others in your field.

Ragland is a self-described results junkie, which was one of the reasons she enjoyed the technology aspect of her early career. She could solve problems and see outcomes through the use of technology that changed how her company worked, and, as a result, she knew she was making a difference.

## Finding A Future in Tech through Mentors and Role Models

Although IT wasn't her original career plan, she soon recognized it as a viable path for her future. While working in the male-dominated consulting industry, she had several mentors who encouraged her to pursue a career in technology. They guided her in addressing the challenges of being one of the few women on the team. As a result, she felt empowered, knowing she had supportive advocates who believed in her.

In her IT career, she worked for women who excelled at articulating their vision for the organization. She believed that was her strength as well, and it helped her see that you could build a career based on your ability to communicate the capabilities and achievements of the IT organization. These role models helped her realize that she, too, could pursue a career in technology.

As she advanced within the IT organization, she began to consider management roles as her ultimate career path. She recognized that she had a vision and a voice to shape the organization and believed she could make a more significant impact as a leader. Encouraged by her peers and senior leaders, she felt inspired to pursue roles with greater responsibility. Her history of having supportive and inspiring leaders throughout her career served as a foundation for her to want to offer the same encouragement and mentorship to others who aspire to a career in technology.

## Heeding the Call to Help Others

This foundation of mentorship and support would prove crucial when Ragland discovered an organization that embodied these same principles on a larger scale. At an i.c.stars event in Chicago, program participants showcased their achievements from their four-month cohort. During the session, alumni also spoke about i.c.stars' impact on their personal and professional lives, including how the organization enabled them to achieve financial independence and break the cycle of generational poverty.

Ragland learned that the program's success stems from recruiting underserved adults who have demonstrated resilience, overcome adversity, are willing to work long hours, and have a passion for technology. All prospective interns must complete a rigorous questionnaire, several interviews, and assessment testing before being accepted into the program.

Another way i.c.stars screens for resiliency is by seeking candidates with a history of working 40 or more hours a week or who possess a bachelor's degree in a non-technical field. In both instances, these candidates have demonstrated dedication to hard work and a commitment to completing a learning curriculum.

Ragland knew she wanted to get involved with i.c.stars because of their track record in identifying and developing the talent needed to provide a fresh and diverse perspective in the technology field. Even though a cohort in her hometown of Milwaukee wouldn't start for another year, she left the event knowing that getting involved with i.c.stars would allow her to connect to her community and assist others in achieving their dreams of a career in IT.

Ragland's enthusiasm for the program grew as she learned more about i.c.stars' unique methodology. Unlike traditional training programs, i.c.stars had developed an approach that addressed not just technical skills, but the broader challenges facing career changers.

## The i.c.stars Approach

Before launching i.c.stars in Milwaukee, Ragland discovered that the program's interns learn and gain experience through hands-on training. Throughout the 14-week immersive training program, they build applications to address real business challenges faced by sponsoring organizations.

She also found that another unique aspect of the i.c.stars program is a wrap-around initiative called BLT (Business, Leadership, and Technology). This serves as an introduction to the business world, which many candidates have never encountered. BLT includes training in conflict resolution, time management, business writing, project management, and PowerPoint skills. Ragland notes, "It helps them understand what it's like to arrive every morning at 8:00 a.m. and be available for meetings. It provides a holistic education that encompasses more than just technical training."

With this thorough understanding of the i.c.stars model, Ragland was ready to help ensure Milwaukee's launch would be just as successful.

## Start-up in Milwaukee

The success of launching a program like i.c.stars in a new city depends on three factors. First, you need staff members who effectively engage with the community and are experienced in operating talent development programs. Next, you must find the right candidates with a passion for technology and the drive to succeed in the program. The staff at i.c.stars continually communicates the program's value within the community to build trust with candidates so that they will invest in their careers and commit to entering the technology field. Finally, the third key to success is finding the hiring organizations that believe this talent is worth bringing into their organizations. They have leaders like Ragland, who commit to a program that helps interns enter and navigate their cohort journey, being there to hire them when they complete the program.

When i.c.stars launched in Milwaukee in 2018, Ragland joined the Advisory Board to help the organization achieve its mission to prepare "change-driven, inner-city future leaders to develop skills in business and technology for high-level careers in information technology." The i.c.stars program in Milwaukee started with 3 4-month cohorts, with approximately 20 interns in each cohort. However, launching a program and ensuring its success are two different challenges. This challenge required Ragland to move beyond advisory support to direct engagement.

Ragland and her team immediately got involved with one of the first cohorts by serving as clients for a project staffed by program interns. They met with interns weekly to review the project's progress, user experience interviews, and proto-types. As a result, her team was able to assess both the output of the team working on the project and the quality of the interns assigned to their initiative. Like many interns in the i.c.stars program, they were considered for positions at her firm at the end of the cohort.

Another way Ragland and her team engaged with i.c.stars was by meeting the interns at the organization's events, which included a unique networking session called High Tea. During the cohort's intense 12-hour days, the interns take a break at 4 p.m. to meet technology professionals from different disciplines, including cybersecurity, application development, and business analysis. The High Tea is also an opportunity for business leaders and technology executives to share their career stories and meet local talent in the program. The interns' participation in these sessions helps them better understand their career options in IT. It also provides opportunities to network with other profession-als and community leaders, which can benefit them in the future.

Experienced professionals can also assist by mentoring and conducting mock interviews. Ragland notes that getting her team involved in these activities with i.c.stars was easy. They were energized by the opportunity to share their expertise with emerging talent and received a fresh perspective from the interns' out-of-the-box thinking. Engaging team members in the program allows them to connect with something beyond corpo-rate-level community responsibility; it provides employees with personal satisfaction in witnessing and supporting others as they transform their lives.

These touchpoints, from project work to High Tea sessions to mock interviews, provide opportunities for interns to showcase their skills and for employers to find talent. The results demonstrate the effectiveness of this multi-pronged approach.

At the end of the 4-month cohort, the goal is to have each intern placed in a job within 6 months. If they are not employed after they complete the program, they have a 2-year residency that includes a job placement service. According to the i.c.stars website, the paid internship has resulted in a 75% placement rate and an average income increase of 50%.

Many program alumni find jobs with employers in Milwaukee who are eager to hire them for their early career positions. The program's graduates have started their IT careers in various positions, including IT support specialist, quality analyst, cloud automation developer, and project manager. Hiring companies have included United Airlines, Northwestern Mutual, Lockheed Martin, and the Milwaukee Brewers baseball team.

## Finding Time to Impact by Just Saying "Yes"

Given the program's success and her positive experience as advisor and hiring manager, Ragland has become an evangelist for this talent development model.

Her advice to technology leaders considering involvement with programs like i.c.stars is simple: "Just say yes and figure it out." But for those ready to take that leap, the path forward is more straightforward than it might seem. Start by researching similar programs in your community. Organizations like i.c.stars exist in most major cities, often operating under different names but with the same mission of developing nontraditional talent.

The business case is compelling: high placement rates, substantial income increases, and access to motivated professionals who bring fresh perspectives on longstanding challenges. More importantly, as Ragland discovered, you gain something unexpected: a renewed sense of purpose in your leadership role.

"You get exposure to people who are changing their lives, their communities, and contributing to companies in ways that refresh and challenge our thinking," she reflects.

In a profession constantly seeking the next innovation, perhaps the most transformative change isn't in our technology, it's in who we invite to help build it.

## CHAPTER 13

# Doing the Next Right Thing: Unlocking Opportunities with Tech Apprenticeships

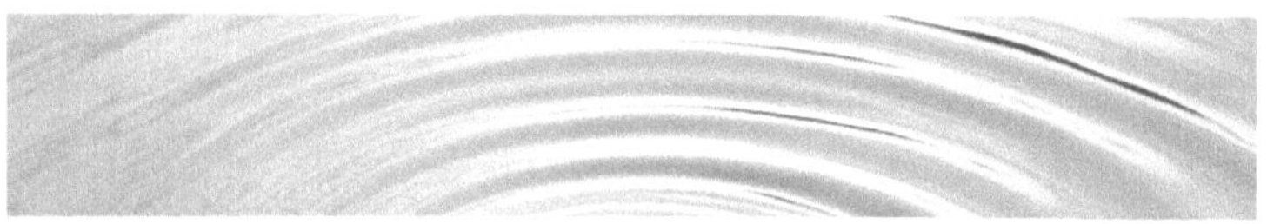

*"This is your second chance at a first career."*
**—Barry Lowry**

The room fell silent as Barry Lowry finished his pitch to Belfast's tech industry leaders. He needed commitments for 30 apprentices to make his program viable, but he'd only secured government backing for 10. Then the Fujitsu Head of Human Resources stood up: "I've known Barry for years, and this is a fabulous idea. Fujitsu will take four apprentices." By the end of that breakfast meeting, they had 32 commitments, and Northern Ireland had its first ICT apprenticeship program.

The success of that morning didn't just mark a turning point for aspiring apprentices; it reflected Lowry's own journey, shaped by unexpected opportunities and a deep commitment to helping others find their path.

Lowry, now a CIO for the Irish government, has taken an unconventional route to becoming a technology leader. Although he never planned to enter the computing field, the opportunities he encountered early in his career, along with the mentors and leaders who recognized and supported his potential, helped Lowry develop into the leader he has become. These formative experiences motivated him to help others by supporting young people in his community, transforming their lives through his roles in the UK and Irish governments.

## Beginning the Journey to a Career in Technology

When Lowry considered attending college, he thought he might pursue a career in teaching or journalism. He received admission offers from Queen's University, the University of Ulster, and the University of Sheffield to study history and politics. Before college began, he took a summer civil service job with the Northern Ireland government. When his employer saw his impressive Advanced Level (A-level) exam results, which are required for college entry in the UK, they were so impressed that they offered him an opportunity to interview for a permanent position.

As he considered the job offer that would require him to postpone his entrance to university, he reflected that many of his summer coworkers were unemployed teachers. He began questioning the value of attending university for three to four years to become a teacher, only to end up jobless. He wondered if he should take the permanent government job for a few years and wait until the teaching job market improved. His mother was concerned about his decision, but his pragmatic father said, "He might be onto something. Let him take a year off and experience work."

Lowry's first job with the government was assisting senior citizens with their pensions. He loved his job and his co-workers, and he especially enjoyed having the opportunity to help residents of Northern Ireland with their pension claims. While in this role, he had a mentor, Paul Hughes, who encouraged him to apply for government training to pursue a career in IT. He asked him why he should apply for this position when he loved his current job. Hughes told him he loved his job because he was 18 years old and might be married with children in a few years, and he wouldn't want (nor could he afford) to work at his current job. Lowry agreed to fill out the form and took the IT aptitude test, resulting in a successful interview and acceptance as a COBOL programmer trainee.

His first supervisor in IT was Jenny Johnston, an exceptional leader who knew how to develop her team members. She told Lowry that he was a good programmer, but she observed him interacting with users and noticed that he instinctively bonded with them, and in return, they trusted him. She recognized that he was uniquely able to understand the users' needs and deliver their desired solutions. As a result, she encouraged him to consider becoming an analyst. He followed her advice and soon discovered that it was his calling.

Johnston continued challenging him to tackle more advanced assignments to help him develop his skills. With her help, he constantly explored the next step on the career ladder to see if he could take on an expanded role. He quickly progressed from junior to senior analyst positions and eventually became a team manager. Lowry notes that he never felt like his career had a ceiling; however, he also never imagined that an 18-year-old from East Belfast could progress to become a government CIO.

Another manager, Anne Conaty, told him he was "very quiet and should be more communicative." She saw that he had the potential to progress to higher levels in the organization and that if he didn't speak up, senior leaders wouldn't notice him, which could result in him being overlooked for new roles. She helped him improve his presentation skills by giving him opportunities to present whenever possible. With her support, his communication skills went from a weakness to a strength, contributing to his career success.

By this stage, Lowry had finally returned to study and was attending Queen's University two nights a week. But rather than take a formal degree in IT, he was more interested at this stage in the components of change and focused on Business Studies. He then completed a Master's in Organisation and Management Theory. As he took on positions with larger teams and increased responsibility, he realized that the theory he had learned (Maslow, Herzberg, and McGregor) could be made true in practice, and if he treated his team members the way Johnston and Conaty treated him, they could achieve remarkable results. He discovered that by treating people this way, you might be disappointed one out of ten times, but you will be absolutely overjoyed nine out of ten.

While focused on advancing his professional career, Lowry also wanted to work with the young people in his community. Now living by the North Down Coast in Bangor, he visited his local church one day and asked if he could volunteer to help with their youth group. Soon after he began to help them, he became the group leader. Working with these young people, he quickly learned that this area of Northern Ireland didn't suffer from material poverty; it faced a poverty of hope and love. Although it was a relatively affluent community, many young people lacked guidance from their parents and had limited ambition. They had no idea what careers they wanted to pursue, and many had been given very little support from either schools or families.

He wanted to find a way to help them pursue careers in technology, just as the mentors and leaders who supported him throughout his career journey had done. Lowry envisioned a formal Information and Communication Technology (ICT) apprenticeship program that would provide training and support for these young people to have IT careers. Now that he knew what he wanted to do, he needed to find a way to gain the support of government leaders to make it a reality.

## The Courage to Do the Next Right Thing

Lowry continued to advance his career in Northern Ireland's government and was asked to lead a team responsible for developing a shared service center for IT and other governmental functions. Governments worldwide have attempted similar initiatives without success; however, Lowry and his colleagues, driven to succeed, established a world-class shared service recognized as a "monumental success." This accomplishment subsequently attracted the attention of government ministers, creating an opportunity for Lowry to engage with them in support of his idea for an IT apprenticeship program.

Now in charge of Government ICT and the Head of the ICT profession, Lowry regularly met with Sammy Wilson, his government minister, and realized Wilson might be able to assist him in achieving his vision. Wilson was an ex-schoolteacher who publicly stated that he wanted the government to consider apprenticeship programs for hiring and developing civil servants.

Following a meeting with the minister, Lowry approached him and asked if he would entertain creating an apprenticeship program for the ICT department he was responsible for. The minister responded, "That is a brilliant idea, and we're going to do it. If you need anything from me, give me a shout."

Lowry later updated Wilson that he had tried to pursue the idea with officials from the Department of Employment and Learning, but they rejected it, stating it wouldn't work. The minister looked at his watch and said, "Let me guarantee you that before the close of business today, you will have a meeting about your idea for an ICT apprenticeship scheme, and it won't be dismissed this time." Wilson phoned the minister responsible for Employment and Learning and arranged the meeting with Lowry. That meeting and the talent and enthusiasm of Lowry's ICT Personnel team eventually led to the creation of the first ICT apprenticeship program in Northern Ireland.

Lowry knew he needed assistance from outside the government to make this happen. He could commit to sponsoring ten people for the first cohort, but he told the minister they should recruit at least thirty. To achieve that, he would need to contact his network in the technology industry to solicit their support. Lowry told them they would be invited to a breakfast meeting hosted by the Minister of Employment and Learning, Stephen Farry, who supported the concept. During that meeting, the minister would ask them to sign up for a new ICT apprenticeship program with the government, saying that "they would need to agree to participate." One of these attendees said, "When Barry phones you up and asks you to do something, he doesn't expect you to say no."

The meeting was arranged, and several senior leaders from local companies were present. Belfast Metropolitan College was engaged to conduct the on-the-job apprentice training and presented its approach to the attendees. The idea was so compelling that the senior recruiter from Fujitsu stood up and said, "I've known Barry for years, and this is a fabulous idea. Fujitsu will take four apprentices." Others followed, and at the end of the meeting, a commitment was made to 32 ICT apprentices.

More than ten years later, the IT apprenticeship program thrives in Northern Ireland, even though Lowry has moved on to other positions. The creation and growth of the program would not have been possible if Lowry hadn't been determined to give back and had the courage to approach the government minister with his idea to address the employment needs in his community. Lowry learned from this experience that politicians want to make a difference in people's lives, but they need the help of equally like-minded civil and public servants.

## Continuing to Pay It Forward with Fastrack into Information Technology (FIT)

In 2016, Lowry relocated to Dublin and assumed the role of CIO for the Government of Ireland. One of his first meetings in his new role was with Peter Davitt, the CEO of Fastrack into Information Technology (FIT). Lowry and Davitt had met years ago in Northern Ireland and had much in common.

Like Lowry, Davitt didn't attend university to study technology. He was an apprentice bricklayer because it was considered a secure profession, and he needed to be able to help support his family. While attending a technical school, he met a chaplain who encouraged him to get involved with the disadvantaged in Ballymun, a suburb of Dublin. He assisted with the youth clubs in the area in the evenings and on weekends, supporting their mission to keep the local youth occupied, thereby preventing them from getting into trouble.

Davitt recognized that a lack of job opportunities was the underlying reason for keeping these young people busy. 66% of the population experienced long-term unemployment, and more than 80% depended on social welfare. To address this, he and others set up a job center to help the residents find employment, only to discover that they lacked the skills that were in demand by employers.

One day, a representative from Microsoft contacted the job center and encouraged them to tell employers they should hire candidates with Microsoft certifications. They didn't know what a Microsoft certification was, but they quickly realized that job applicants with this accreditation were immediately employable. The issue was that the certification course cost approximately 1,000 euros, which was out of reach for most applicants. To address this, the job center secured the necessary funding to provide a certification course for 25 individuals who had been long-term unemployed. All 25 completed the course, received their certification, and secured jobs. They repeated this success story multiple times, building trust in their ability to help residents secure work in the technology profession.

During this period, the tech industry in Ireland experienced rapid growth; however, the country faced a shortage of skilled resources. Davitt and his colleagues approached Ireland's Industrial Development Authority (IDA) to tell them they could help solve the problem. With more than 250,000 long-term unemployed individuals, they informed the IDA that they could help provide training to address the technology skills gap. The IDA organized a group of tech companies, including IBM and Microsoft, to listen to the proposal. They were intrigued by the idea and successfully lobbied the government to fund the program for the next three years, and with that, FIT was born.

When Davitt met with Lowry in 2016, he had hoped to persuade the Irish government to engage the services of FIT. In the middle of Davitt's pitch, Lowry stopped him and asked, "Are you suggesting that the government might establish an apprenticeship scheme with you?" Davitt said, "Yes, that's precisely it." Lowry responded," The answer is yes, so let's figure out how to make that happen."

Lowry engaged his peer network in Ireland to support the FIT apprenticeship program, including CIOs and senior executives from companies such as Boston Scientific, JP Morgan, and Aer Lingus. This group supported apprentices, and several members are now on the FIT Board of Directors. These employers were attracted to the FIT Tech Apprenticeship program for its access to the most in-demand talent and skills, including software development, computer networking, and cybersecurity.

An apprentice in the program spends the first six months of the two-year cycle in a concentrated series of courses to learn the necessary technical skills. Following that, they spend the next 18 months gaining on-the-job experience while continuing college-level training. Once an apprentice completes the two-year cycle, approximately 95% of their employers retain them for continued employment.

By the end of 2024, FIT had trained over 800 tech apprentices and engaged more than 230 employers. In the first two cycles (a total of four years), under Lowry's leadership, the Irish government had more than 180 apprentices who had, or are working to become civil servants working in ICT. Lowry and the Irish government's support for FIT has encouraged other organizations in Ireland to get involved. Davitt notes, "Without him [Lowry], we wouldn't have gotten the traction. He's a champion for what we're about and championed it right through the public sector."

## Each Successful FIT Apprentice Is a Story of Transformation

One of the earliest apprentices in the FIT program was a young man training to become a butcher. He was married with two young children when he lost his job. He looked in a local paper to find the jobs that paid the most and determined they were in computer programming. That is when he decided to pursue a career in technology.

He immediately contacted the job center and became an apprentice in the program. Following his apprenticeship, he advanced his career to a senior leadership role and is now a FIT Board of Directors member. Davitt observes that many successful FIT apprentices, including this former butcher, come from backgrounds where opportunities are scarce. However, when given the chance to become a FIT apprentice, they appreciate it and maximize their potential.

Like Davitt, Lowry understands these apprentices "have a natural hunger, and they see an opportunity that they didn't think was there before, and they want to seize it." He believes the FIT apprenticeship program offers a "second chance at a first career" and provides access to a profession with almost limitless opportunities for advancement.

Apprentices on Lowry's teams can apply their technology skills to drive innovation in government services for the Irish people. One example is the award-winning website whereyourmoneygoes.gov.ie. The idea for the site was to provide a simple and easy-to-use tool that explains to Irish citizens how the government develops its budget. Lowry employed his data analytics and web teams to create this site, each of which included an apprentice. When launched, the site was recognized with an award for contributing to governmental fiscal awareness. Lowry explained to his team that the award was for more than the website; it recognized their contributions to their country.

Another application that the apprentices on his team had the opportunity to contribute to was developed during the COVID-19 pandemic. The Irish government recognized that once vaccinations were initiated, people would eventually need a credential to prove they had received the vaccine. Lowry's team, including apprentices, developed an application in six weeks that enabled Irish citizens to provide digital proof of vaccination. This digital credential was used for travel and hospitality purposes. It allowed people to enter coffee shops and restaurants by providing digital verification that everyone visiting these establishments had received the vaccine.

It was a political win and another opportunity for Lowry to recognize his team's contribution. He told them they should go home and inform their parents that they were part of the team that built the app that helped safeguard people's health across Ireland. He notes that government technology jobs will typically pay less than those in other industries, but these roles offer a unique opportunity to make a wide-reaching impact that other sectors can't match.

## The Privilege of Being a Technology Leader

Lowry acknowledges he has been privileged to have the career he's enjoyed. He knows that when you reach a certain level in a profession like IT, you attract attention and interest from others. When that happens, you should ask yourself what to do with this influence. Like Lowry, it could be as simple as making phone calls to organize your peers to support apprenticeship programs, such as FIT. He believes that was the least he could do with the influence that came with his position.

We all may wonder how we can make a difference in the world. Lowry tells many people he mentors, "Know where your heart is and follow your heart, but keep your head not too far after it. If your heart tells you to do things for others, you will find ways to make it happen." By following his heart, Lowry has helped transform the lives of many young people across Ireland who have benefited from the apprenticeship programs he has supported throughout his career.

## CHAPTER 14

# Impacting the Profession: What Did We Learn

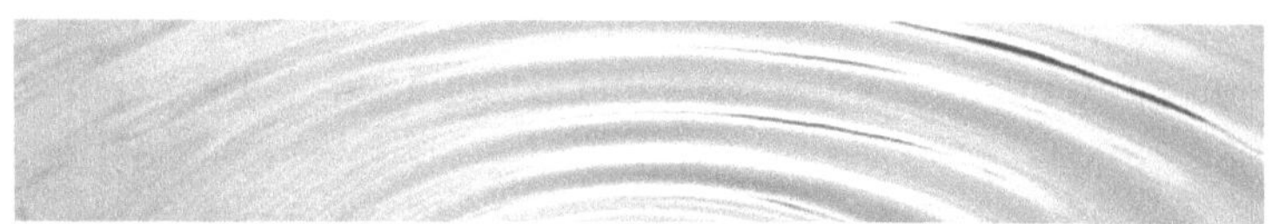

## Your Impact Starts Now: Inspiring the Next Tech Generation

The statistics presented in this section paint a stark picture: despite decades of technological progress, the computing field still faces challenges with diversity and representation. Women comprised just 22% of U.S. computer science graduates in 2022, still below the 37% peak in 1985. The participation of underrepresented groups in the technology profession continues to fall short of their population share. These figures are more than just statistical gaps; they limit the profession's future potential and capacity for innovation.

Yet within these challenges lies an extraordinary opportunity for today's technology leaders. The leaders profiled in this section prove that individual commitment can create systemic change, one mentorship, one program, one courageous action at a time.

## The Business Case for a Deeper
## and More Diverse Talent Pool

Expanding and diversifying the technology workforce is essential for business success. 75% of survey respondents who identified school-age and college students as top priorities for building the talent pipeline understand a key truth: the challenges of tomorrow will need perspectives and solutions that only come from diverse teams representing the communities they serve.

Dr. Philile Mkhize's journey from a rural South African village to becoming the COO of Technology at a pan-African bank illustrates what's possible when potential is recognized and nurtured. Her work with Technovation, mentoring young women to use technology to solve real-world problems like gender-based violence, shows how personal transformation can create ripple effects that extend well beyond individual success stories.

Similarly, the comprehensive approach of STEM Advantage, co-founded by Cindy McKenzie, demonstrates how offering students a support network during their university years can lead to impressive outcomes. Their 100% graduation rate among participants, far exceeding national averages, proves that success is possible with the right combination of mentorship, internships, and scholarships.

## The Power of Perspective and Purpose

What distinguishes the leaders in this section isn't their titles or budgets; it's their perspective on what technology leadership means. Barry Lowry didn't just see government IT as maintaining systems; he saw it as creating pathways for young people to transform their lives. When he tells apprentices that their work on Ireland's COVID vaccination app helped "safeguard people's health across Ireland," he's communicating to connect their daily tasks to a larger purpose.

This perspective shift, from managing technology to developing people, represents a fundamental evolution in technology leadership. It recognizes that our industry's most significant challenges aren't technical but human: How do we ensure that tomorrow's technology workforce reflects the diversity of the world it serves? How do we create pathways for talent that exist outside traditional pipelines? How do we use our platforms and influence to lift others?

## The Multiplying Effect of Individual Action

This section's stories demonstrate how individual actions can generate far-reaching ripple effects. Simon Reiter's commitment to mentoring women annually through Women 4 STEM doesn't just impact those mentees; it transforms how he leads, making him more inclusive in hiring and effective in leading his teams. His leadership team evolved from exclusively male to 50% female, showing how external mentoring can drive internal change.

Julie Ragland's involvement with i.c.stars in Milwaukee illustrates how saying "yes" to one opportunity creates cascading benefits. Her organization gained access to diverse talent with fresh perspectives while providing life-changing opportunities for program participants. The 75% placement rate and 50% average income increase for i.c.stars graduates represent more than career advancement; they represent breaking generational poverty cycles and creating new possibilities for entire families and communities.

## The Strategic Approach to Systemic Change

What emerges from these stories is a framework for creating last-ing change:

**Start Early, Think Long-Term:** Dr. Mkhize's work with Tech-novation focuses on pre-college girls, recognizing that early exposure to technology's potential for social impact can influ-ence career aspirations before traditional barriers become established.

**Create Multiple Entry Points:** The i.c.stars model demonstrates how adult career changers can successfully enter technology, while FIT's apprenticeship program provides an alternative to traditional university pathways. Different people need different on-ramps into the profession.

**Combine Individual Support with Systemic Solutions**: STEM Advantage doesn't just provide scholarships; it creates compre-hensive support systems, including mentorship, internships, and professional development.

**Leverage Your Platform for Others**: Barry Lowry used his posi-tion as government CIO to create apprenticeship programs and advocate for policy changes. Technology leaders can exert their influence and credibility, opening doors for others.

## The 99% Who See This as Essential Work

A key takeaway from the survey in this section is that 99% of global technology leaders consider expanding and diversifying the tech talent pool a top priority, while only 1% believe it's unnec-essary. This represents universal alignment on the importance of this mission. The challenge isn't convincing leaders that this work matters; it's helping them find their path to contribution.

The leaders profiled here offer multiple models: Dr. Mkhize's global mentoring and judging work, Cindy McKenzie's comprehensive organizational building, Simon Reiter's focused annual mentoring commitment, Julie Ragland's strategic advisory board participation, and Barry Lowry's policy advocacy and program creation. There's no single "right" way to make an impact; there are only countless opportunities for leaders to find approaches that align with their passions, skills, and circumstances.

## The Time for Action

As you consider your role in shaping the profession's future, remember that the question isn't whether to get involved; it's how to get involved to maximize your impact and align with your values. Every leader featured in this section started by saying "yes" to an opportunity to help others.

The technology profession is entering a new era. Advances in artificial intelligence, quantum computing, and other emerging technologies are reshaping how we live and work. To thoughtfully and ethically fulfill the promise of these new technologies, we will need talented and diverse teams that understand the needs of the firms and communities they serve.

The leaders who recognize this moment and act on it will shape the trajectory of the entire profession. Whether mentoring one person, creating programs that impact thousands, changing hiring practices, or advocating for policy reforms, every technology leader can be part of the solution.

The question isn't whether you have enough time, resources, or influence to make an impact. The question is whether you'll use what you have to expand opportunity for others. As Barry

Lowry advises his mentees, "Know where your heart is and follow your heart, but keep your head not too far after it. If your heart tells you to do things for others, you will find ways to make it happen."

The profession's future depends not on the next breakthrough technology but on the next generation of diverse talent we choose to develop today. That future begins with your next decision to invest in someone else's potential.

In the words of Dr. Mkhize, "The work starts today."

## Impacting the Profession: A Final Reflection

While recovering from foot surgery, confined to a cast and crutches for two weeks, I found myself with something I hadn't had in years: time to think. My mind drifted to my bucket list and what I would do once I could walk again. As I mentally checked off dreams of travel and once-in-a-lifetime experiences, an uncomfortable truth surfaced: every single item was about me and for me. What if I'd been measuring my life by the wrong list? What if the most meaningful goals weren't about what I could do for myself but about what I could do for and with others?

Shortly after I returned to work, an opportunity arose that would change my bucket list. I learned that the local Boys & Girls Clubs needed volunteers for a three-week artificial intelligence summer camp designed to introduce high school students from underrepresented groups to technology careers. At my next town hall meeting, I shared the opportunity with my IT team, uncertain if we'd find the 60 volunteers needed. By the time I walked back to my desk, emails were already flooding in; team members volunteering without hesitation, even before knowing exactly what would be asked of them. More than seventy people stepped up. After months of planning, the camp was successfully launched across multiple sites. On the final day, as I stood on stage presenting awards alongside the head of the Boys & Girls Clubs, it hit me: together, we had accomplished something that mattered far more than anything on my original bucket list.

Nearly three years later, well into my retirement, I met Paige Frank, a student from that first camp, who had just been named Pennsylvania Youth of the Year. Before attending camp, Paige had planned to go to cosmetology school; technology wasn't on her radar. After three weeks at camp, she left inspired to pursue a career in tech, returned as a peer instructor, and went on to study information science on a scholarship. That encounter reminded me that the most meaningful bucket list items aren't about what we do for ourselves but what we do for others. When we align our skills with a greater purpose, we don't just serve our communities; we strengthen our people and change lives in the process.

# The Interconnected Journey of Technology Leadership

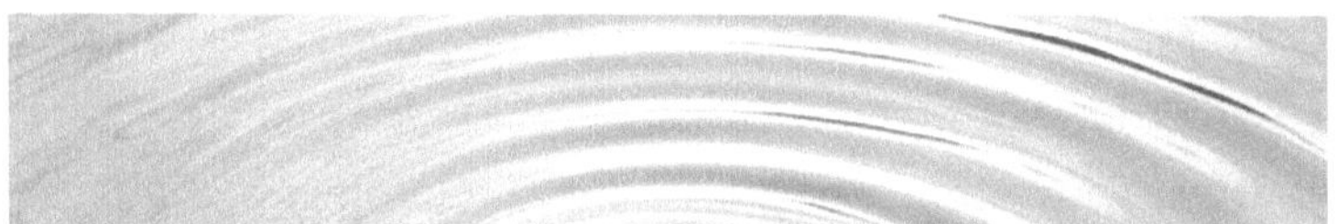

Extensive research, including surveys of 1,600 technology leaders across more than 60 countries and in-depth interviews with successful practitioners, reveals three core areas that define outstanding technology leadership: driving innovation within organizations, motivating and developing teams, and inspiring and advancing the broader technology profession. The three impact areas covered in this book: your firm, your team, and the profession, are interconnected parts of a unified leadership strategy. While each is distinct, they share several key principles and practices that transcend the boundaries of each impact area.

This interconnectedness means your journey as a technology leader involves making simultaneous investments in all three areas. You can't sustain innovation without an engaged team. You can't build a thriving workplace without linking it to meaningful business outcomes. And you can't create lasting change in your firm or team without contributing to the growth of the profession that makes it all possible.

While these three impact areas require different strategies and tactics, they all rest on a foundation of shared leadership principles that consistently appear across the most successful technology organizations.

## Principle #1: Prioritize the People

The most impactful technology leaders understand that sustainable success requires putting people first, recognizing their unique needs, aspirations, and learning styles rather than forcing them into rigid organizational structures. Steve Zerby at Owens Corning exemplified this philosophy through his core belief that "you need to build jobs around people instead of putting people in jobs." Rather than following formulaic approaches to structuring IT organizations, Zerby mixed and matched roles to meet both business needs and individual development goals, creating positions that allowed high-potential employees to gain diverse experiences while maintaining continuity in business relationships.

Cindy McKenzie's journey from entertainment industry CIO to STEM Advantage co-founder illustrates how prioritizing people extends beyond current employees to future generations of technology professionals. When she attended Parents Night at UCLA and counted only a handful of women among hundreds in the computer science program, she didn't simply lament the gender disparity; she took action to address it. Her decision to co-found STEM Advantage represented a fundamental shift from managing existing talent to creating pathways for underrepresented groups, recognizing that the profession's future depended on expanding access and opportunity for diverse populations.

These leaders demonstrate that prioritizing people isn't just about accommodating individual preferences; it's about understanding that organizational success comes from aligning systems, processes, and opportunities with human potential. Whether through creating roles that foster career growth, implementing technologies that suit learning styles, or developing programs that broaden access to the profession, these leaders demonstrate that when you genuinely prioritize people, individuals, and organizations flourish in ways that traditional, one-size-fits-all methods can't achieve. However, prioritizing people requires more than good intentions; it demands ensuring that every level of leadership shares this commitment, starting with the quality of direct supervisors.

## Principle #2: Focus on People Development

The most impactful technology leaders know their greatest legacy isn't the systems they build, it's the people they develop. Vishal Gupta's *Focus to Future* sessions pair employees from different departments to strengthen communication and collaboration. Simon Reiter's coaching helped a mentee conquer her fear of public speaking, allowing her to confidently speak in front of hundreds at an industry conference. These leaders show that developing others isn't optional; it's the foundation for lasting organizational impact.

Exceptional leaders see potential where others see limitations, creating growth opportunities that extend beyond immediate business needs. Gupta's F2F format wasn't just about creating innovative ideas; it was a deliberate way to develop cross-functional relationships and communication skills to support employees throughout their careers. Reiter's year-long mentoring relationship didn't just transform his mentee's confidence; it shifted her entire career trajectory and made him a more inclusive and effective leader.

What distinguishes these leaders is their recognition that every interaction is an opportunity to develop someone's potential, every challenge is a chance to build capability, and every success should inspire others to reach higher. Quality supervision, however, goes beyond managing day-to-day performance; the best supervisors actively invest in growing their people's capabilities for the future.

## Principle #3: Direct Involvement Signals True Commitment

The most effective technology leaders don't delegate their key initiatives; they lead by example with personal, hands-on involvement that signals organizational priorities. Jim Rinaldi's annual pilgrimage with his NASA JPL team to the Consumer Electronics Show wasn't just about technology scouting; it was a powerful demonstration that innovation mattered enough for the CIO to personally invest time and resources in discovering emerging trends. Julie Ragland's direct engagement with i.c.stars, from serving on the advisory board to having her team work directly with program participants during High Tea sessions and mock interviews, showed her organization that diversity and talent development weren't just corporate talking points but genuine priorities worthy of executive attention. Dick Daniels' practice of personally visiting hospitals to walk in the shoes of his users sent an unmistakable message to his team: understanding the human impact of their technology work was so important that the CIO himself would take time to experience it firsthand.

This hands-on leadership style creates a ripple effect across the organization, motivating teams to embrace these priorities with the same level of commitment as their leaders. When Rinaldi's team saw their CIO walking the floors of CES year after year, they understood that staying current with emerging technologies and thinking creatively about applications was part of the organization's DNA. Ragland's involvement with i.c.stars energized her team members, who eagerly volunteered to mentor and work with program participants because they saw her genuine passion for the organization. Daniels' hospital visits didn't just reconnect him with the organization's mission; they fostered a culture where his entire IT team recognized the importance of viewing their work through the lens of those they serve. These leaders demonstrate that when executives personally champion what matters most, they don't just convey priorities; they embed them into the organization's culture in ways that last long after they leave the firm.

## The Leadership Qualities That Define Impactful Leaders

Throughout the stories in this book, three leadership qualities stand out as vital for creating meaningful impact: listening to learn, communicating to connect, and acting courageously to do the next right thing. These aren't just nice-to-have soft skills; they are the core capabilities that enable technology leaders to manage the complex human dynamics involved in driving change.

**Listening to Learn** has appeared in every transformation story we've explored. Renee Zaugg's vulnerability-first approach at Otis Elevator, where she shared her personal "entourage tapestry" and asked her team to do the same, created the foundation of trust necessary for leading through crisis. Graeme Thompson's listening tour at Informatica enabled him to understand

and leverage the company's collaborative culture to drive business transformation.

These leaders understand that listening isn't passive. It's an active skill that requires setting aside one's assumptions and agenda to truly understand the perspectives, motivations, and concerns of others. Whether you're trying to drive innovation, increase engagement, or expand the technology talent pool, lasting change begins with understanding the people you're trying to influence.

**Communicating to Connect** transforms individual insights into shared understanding and collective action. The most effective technology leaders don't just share information; they create emotional and intellectual connections that inspire others to act. When Thompson helped his Informatica team understand how their work supporting the company's cloud transformation connected to helping pharmaceutical companies accelerate vaccine development, he wasn't just explaining strategy but connecting daily tasks to life-saving impact.

Dick Daniels' framework of developing unified messaging with his leadership team and then repeating it consistently across the organization demonstrates how communication becomes a tool for alignment and empowerment. The banking CIO shifted her team's mindset from proudly saying they are "good at fixing things when they break" to understanding how system outages negatively affect customers. This change highlights how effective communication can transform understanding and the organization's identity and purpose. It evolved the support team from reactive problem solvers to proactive customer-focused professionals who work to prevent issues instead of just responding to them.

**Acting Courageously to Do The Next Right Thing** separates leaders who talk about change from those who create it. This courage shows up in different ways depending on the situation. Still, it always involves choosing what's right for the long-term success of the organization and the people within it, even when it's hard or unpopular.

Courage takes many forms. For Renee Zaugg, it was the bold choice to invest in infrastructure during a crisis, recognizing it as vital to her organization's future despite the difficult timing. For Steve Zerby and Hans Keller, it was the resolve to turn down candidates with exceptional technical skills whose personalities would undermine the cultures they had carefully built. For Barry Lowry, it was using his platform and influence to open doors for others, leveraging relationships with government ministers and industry leaders to launch apprenticeship programs. In each case, these leaders demonstrated the conviction to act on their principles, even when the path forward was uncertain or required personal risk.

These three qualities: listening to learn, communicating to connect, and acting courageously, work together to amplify the three leadership principles, creating a comprehensive approach to technology leadership that transcends any single initiative or challenge.

## The Choice Before You

This integrated approach to leadership raises an important question about your career journey: How will you apply these insights in your current role? The question isn't whether you have the title, budget, or formal authority to impact these three areas. Every technology professional, from individual contributor to C-suite executive, has opportunities to drive innovation, develop others, and expand access to the profession. The question is whether you'll see and act on those opportunities.

This choice will define not just your effectiveness as a leader but your satisfaction as a professional. When you align your work with a purpose larger than yourself, whether solving meaningful business problems, helping others grow their careers, or opening doors for the next generation, you discover that technology leadership becomes less about managing complexity and more about creating possibility.

The choice is yours. The opportunity is now. And the impact you create will extend far beyond anything you can imagine.

# Being a Technology Leader, It Really Can Be a Wonderful Life

"Strange, isn't it? Each man's life touches so many other lives.
When he isn't around, he leaves an awful hole, doesn't he?"
**—Clarence Odbody, *It's a Wonderful Life* (1946)**

After finishing this book, you might still ask yourself: Why bother having an impact? Why add the pressure of making a difference on top of the stress and tough decisions that already come with leadership? After all, won't I be forgotten once I leave, and will any impact I made fade away with me? Sometimes I wonder if my role even matters to anyone at all.

Those questions are valid. They're the same ones many leaders quietly carry with them every day. But here's the truth: impact isn't about being remembered forever. It's about creating ripples now; ripples that can change someone's career, spark an idea that reshapes a business, or inspire a team to reach higher. You may never see how your influence travels, but it will travel and matter.

It reminds me of the story underlying the classic movie *It's a Wonderful Life*. The film follows the life of George Bailey, played by Jimmy Stewart, who dreams of leaving his hometown of Bedford Falls and becoming an architect, traveling around the world, and building bridges, airfields, and skyscrapers. Every time George is ready to leave Bedford Falls, events prevent him from pursuing his dreams. Regretfully, he put those dreams aside, remained in Bedford Falls for the rest of his life, and struggled financially and emotionally as he watched others successfully pursue their dreams.

George utters in a crisis, "I suppose it would've been better if I had never been born at all. " With those words, the rest of the movie is set into motion. His guardian angel, Clarence Odbody, helps George see how the world would have been had he "never been born at all." This includes how he impacted the lives of the citizens of Bedford Falls, including his younger brother Harry.

When Harry was nine years old, he fell through the ice in a pond and nearly drowned, but George saved his life. Harry eventually joined the Navy during World War II and received the Congressional Medal of Honor for saving the lives of all the soldiers on the transport ship they were traveling on.

George begins to realize that if he had never been born, he wouldn't have been there to save his brother's life, and Harry wouldn't have been there to save all those soldiers.

This same principle plays out in real life in ways we rarely expect, like the day a young man and his fiancée appeared unannounced at his mother's door on a Saturday afternoon. He began the conversation by thanking her for ensuring he learned to swim as a child and then shared the story of what had happened just hours earlier.

While walking through a local park, the couple heard boys playing by a creek. Suddenly, shouts for help erupted when one of the boys, unable to swim, slipped into the water and started drowning. Without hesitation, the young man dove in, pulling the boy to safety.

He realized that his ability to save another life traced directly back to his mother's insistence on swimming lessons years before. And now, because of that single choice she made, another child has the chance to live a full life, one that will create its own ripples for generations to come.

We often don't realize that, as technology leaders, we can influence those around us daily. We may not always see the immediate impact of our actions and question whether leadership's daily struggles are worth it. But, like this mother, our impact may not be realized for many years. As Clarence Odbody said, "Each man's life touches so many other lives."

One last note about this mother and her son: he went on to become an entrepreneur, running several businesses, rose to the rank of captain in the Army National Guard, married his fiancée, and became a father. Many would see him as the hero of this story, and rightly so, but I also see his mother as a hero. She is both a hero and my wife, Janice.

That young man is Michael. And together with his older brother, Patrick, Janice and I are proud to call them our sons.

Now that you've finished this book, ask yourself: How will you create your own ripples in the lives of others by impacting your firm, your team, and the future of our profession?

# Impacting the Firm:
## Innovation Discovery Skills Survey

**Survey Question:**

*"Which of the five discovery skills have you fostered in the organizations you've led to build a culture of innovation?" (Multiple selections allowed)*

- Experimenting
- Questioning
- Observing
- Networking
- Associating

## Methodology

- **Total Responses:** 745 technology leaders
- **Geographic Coverage:** 47 countries
- **Survey Method:** Online questionnaire distributed through LinkedIn
- **Response Rate:** 14%

# Respondent Demographics

**By Geographic Region:**
- U.S.: 51% (381 respondents)
- Non-U.S.: 41% (364 respondents)

**By Organization Size:**
- <1,000 employees: 26%
- 1,000-5,000 employees: 26%
- 5,000-10,000 employees: 13%
- 10,000 employees: 35%

**By Leadership Level:**
- C-Suite (CIO, CTO, CDO): 87%
- Director / Manager: 4%
- CISO: 9%

**By Gender:**
- Male: 76%
- Female: 24%

**By Industry:**
- IT Services and IT Consulting (n=55)
- Financial Services (n=41)
- Government Administration (n=40)
- Hospitals and Health Care (n=39)
- Higher Education (n=51)

# Ranked Results by Respondent Type and Discovery Skill

| Experimenting | |
| --- | --- |
| **Respondent Type** | **Response %** |
| CIO 100 | 93% |
| Financial Services | 88% |
| Government Administration | 88% |
| IT Services and IT Consulting | 86% |
| Firms with more than 10,000 employees | 84% |
| CxO / VP IT / Head IT | 80% |
| Male | 78% |
| U.S. | 78% |
| Non-U.S. | 78% |
| Firms with less than 1,000 employees | 77% |
| Female | 77% |
| Firms with 1,000-5,000 employees | 76% |
| Hospitals and Health Care | 74% |
| Firms with 5,000-10,000 employees | 73% |
| CISO | 71% |
| Higher Education | 69% |
| Director / Manager | 61% |

| Questioning | |
| --- | --- |
| **Respondent Type** | **Response %** |
| CIO 100 | 94% |
| IT Services and IT Consulting | 93% |
| Government Administration | 93% |
| Firms with 5,000-10,000 employees | 91% |
| Male | 90% |
| Financial Services | 90% |
| Hospitals and Health Care | 90% |
| Firms with more than 10,000 employees | 89% |
| CISO | 89% |
| Firms with 1,000-5,000 employees | 89% |
| U.S. | 89% |
| CxO / VP IT / Head IT | 89% |
| Firms with less than 1,000 employees | 89% |
| Non-U.S. | 89% |
| Director / Manager | 87% |
| Female | 85% |
| Higher Education | 82% |

## Observing

| Respondent Type | Response % |
| --- | --- |
| Financial Services | 95% |
| CIO 100 | 83% |
| Government Administration | 80% |
| Firms with more than 10,000 employees | 80% |
| Firms with 5,000-10,000 employees | 78% |
| CxO / VP IT / Head IT | 78% |
| Firms with 1,000-5,000 employees | 78% |
| Male | 77% |
| Female | 77% |
| Hospitals and Health Care | 77% |
| Non-U.S. | 77% |
| U.S. | 76% |
| IT Services and IT Consulting | 75% |
| Firms with less than 1,000 employees | 74% |
| CISO | 71% |
| Director / Manager | 65% |
| Higher Education | 63% |

| Networking | |
| --- | --- |
| **Respondent Type** | **Response %** |
| CISO | 85% |
| Female | 82% |
| Higher Education | 82% |
| Government Administration | 80% |
| Firms with more than 10,000 employees | 78% |
| IT Services and IT Consulting | 78% |
| Non-U.S. | 76% |
| CxO / VP IT / Head IT | 75% |
| U.S. | 75% |
| Firms with 5,000-10,000 employees | 75% |
| CIO 100 | 74% |
| Firms with 1,000-5,000 employees | 73% |
| Male | 73% |
| Firms with less than 1,000 employees | 71% |
| Director / Manager | 68% |
| Financial Services | 68% |
| Hospitals and Health Care | 67% |

| **Associating** | |
| --- | --- |
| **Respondent Type** | **Response %** |
| CIO 100 | 74% |
| Government Administration | 70% |
| Financial Services | 66% |
| Firms with 5,000-10,000 employees | 66% |
| Firms with over 10,000 employees | 66% |
| U.S. | 65% |
| Higher Education | 65% |
| IT Services and IT Consulting | 65% |
| Firms with less than 1,000 employees | 63% |
| Female | 63% |
| Male | 63% |
| CxO / VP IT / Head IT | 63% |
| Hospitals and Health Care | 62% |
| Non-U.S. | 62% |
| CISO | 62% |
| Firms with 1,000-5,000 employees | 60% |
| Director / Manager | 58% |

# Impacting the Team:
## Herzberg's Two-Factor Theory Survey

**Survey Question:**

*"What are the five most important areas that you focus on to create a great workplace?" (Multiple selections allowed)*

- Salaries and benefits
- Quality of supervisors
- Company policies and procedures
- Positive coworker relationships
- Working conditions
- Employee development
- Assigning meaningful work to employees
- Expanding employee job responsibilities
- Recognition
- Employee career advancement

## Methodology
- **Total Responses:** 815 technology leaders
- **Geographic Coverage:** 53 countries

- **Survey Method:** Online questionnaire distributed through LinkedIn
- **Response Rate:** 25%

## Respondent Demographics

### By Geographic Region:

- U.S.: 53% (432 respondents)
- Non-U.S.: 47% (383 respondents)

### By Organization Size:

- <1,000 employees: 25%
- 1,000-5,000 employees: 25%
- 5,000-10,000 employees: 13%
- 10,000 employees: 37%

### By Leadership Level:

- C-Suite (CIO, CTO, CDO): 83%
- Director / Manager: 4%
- CISO: 13%

### By Gender:

- Male: 67%
- Female: 33%

### By Industry:

- IT Services and IT Consulting (n=32)
- Financial Services (n=40)
- Government Administration (n=36)
- Hospitals and Health Care (n=56)
- Higher Education (n=45)

# Ranked Results by
# Respondent Type and Hygiene Factor

## Salaries and Benefits

| Respondent Type | Response % |
|---|---|
| Director / Manager | 82% |
| Best Places | 79% |
| CISO | 75% |
| Financial Services | 71% |
| Male | 70% |
| Firms with more than 10,000 employees | 70% |
| Firms with 5,000 to 10,000 employees | 69% |
| Non-U.S. | 69% |
| Firms with 1,000 to 4,999 employees | 67% |
| U.S. | 65% |
| CxO / VP IT / Head of IT | 65% |
| Higher Education | 64% |
| Firms with less than 1,000 employees | 64% |
| IT Services and IT Consulting | 63% |
| Female | 61% |
| Hospitals and Healthcare | 59% |
| Government Administration | 56% |

## Quality of Supervisor

| Respondent Type | Response % |
| --- | --- |
| Best Places | 64% |
| Financial Services | 63% |
| IT Services and IT Consulting | 63% |
| U.S. | 57% |
| Firms with 5,000 to 10,000 employees | 54% |
| Government Administration | 53% |
| Hospitals and Healthcare | 53% |
| CxO / VP IT / Head of IT | 52% |
| Firms with 1,000 to 4,999 employees | 52% |
| Firms with less than 1,000 employees | 51% |
| Male | 51% |
| Firms with more than 10,000 employees | 50% |
| Female | 50% |
| Director / Manager | 50% |
| CISO | 49% |
| Higher Education | 48% |
| Non-U.S. | 45% |

## Company Policies & Procedures

| Respondent Type | Response % |
| --- | --- |
| Firms with less than 1,000 employees | 11% |
| Best Places | 10% |
| Female | 10% |
| IT Services and IT Consulting | 9% |
| Non-U.S. | 9% |
| CxO / VP IT / Head of IT | 8% |
| Financial Services | 8% |
| Director / Manager | 7% |
| Firms with 1,000 to 4,999 employees | 7% |
| Firms with more than 10,000 employees | 7% |
| Male | 6% |
| Firms with 5,000 to 10,000 employees | 6% |
| U.S. | 6% |
| Hospitals and Healthcare | 6% |
| Government Administration | 6% |
| CISO | 4% |
| Higher Education | 2% |

# Positive Coworker Relationships

| Respondent Type | Response % |
| --- | --- |
| Higher Education | 68% |
| Director / Manager | 61% |
| Firms with less than 1,000 employees | 59% |
| Female | 56% |
| IT Services and IT Consulting | 56% |
| Non-U.S. | 54% |
| CISO | 54% |
| Government Administration | 53% |
| Firms with 1,000 to 4,999 employees | 52% |
| CxO / VP IT / Head of IT | 52% |
| Male | 51% |
| U.S. | 51% |
| Best Places | 50% |
| Firms with 5,000 to 10,000 employees | 47% |
| Hospitals and Healthcare | 47% |
| Firms with more than 10,000 employees | 46% |
| Financial Services | 37% |

# Working Conditions

| Respondent Type | Response % |
| --- | --- |
| Government Administration | 44% |
| Non-U.S. | 42% |
| Firms with 5,000 to 10,000 employees | 40% |
| Hospitals and Healthcare | 37% |
| CISO | 37% |
| Firms with less than 1,000 employees | 37% |
| Director / Manager | 36% |
| Male | 34% |
| Financial Services | 34% |
| Female | 34% |
| IT Services and IT Consulting | 34% |
| CxO / VP IT / Head of IT | 34% |
| Firms with more than 10,000 employees | 33% |
| Firms with 1,000 to 4,999 employees | 31% |
| Best Places | 30% |
| U.S. | 26% |
| Higher Education | 25% |

## Ranked Results by Respondent Type and Motivator

| Employee Development | |
| --- | --- |
| **Respondent Type** | **Response %** |
| IT Services and IT Consulting | 84% |
| Higher Education | 80% |
| Government Administration | 78% |
| Firms with 5,000 to 10,000 employees | 72% |
| CISO | 72% |
| Best Places | 70% |
| Non-U.S. | 69% |
| Firms with less than 1,000 employees | 69% |
| Female | 69% |
| Hospitals and Healthcare | 69% |
| Firms with more than 10,000 employees | 68% |
| CxO / VP IT / Head of IT | 68% |
| Male | 68% |
| U.S. | 67% |
| Firms with 1,000 to 4,999 employees | 65% |
| Director / Manager | 61% |
| Financial Services | 61% |

| Assigning Meaningful Work to Employees | |
| --- | --- |
| **Respondent Type** | **Response %** |
| Director / Manager | 75% |
| U.S. | 74% |
| Firms with 1,000 to 4,999 employees | 74% |
| Firms with more than 10,000 employees | 74% |
| Female | 74% |
| Financial Services | 71% |
| CxO / VP IT / Head of IT | 71% |
| Government Administration | 69% |
| Male | 69% |
| Hospitals and Healthcare | 69% |
| CISO | 68% |
| Non-U.S. | 67% |
| Firms with less than 1,000 employees | 66% |
| Higher Education | 66% |
| Firms with 5,000 to 10,000 employees | 64% |
| Best Places | 61% |
| IT Services and IT Consulting | 53% |

| Expanding Employee Job Responsibilities | |
| --- | --- |
| **Respondent Type** | **Response %** |
| CISO | 27% |
| Financial Services | 26% |
| Male | 26% |
| Firms with 1,000 to 4,999 employees | 25% |
| Higher Education | 25% |
| Best Places | 24% |
| U.S. | 24% |
| Firms with less than 1,000 employees | 24% |
| Firms with more than 10,000 employees | 23% |
| CxO / VP IT / Head of IT | 23% |
| Non-U.S. | 23% |
| Hospitals and Healthcare | 22% |
| Government Administration | 19% |
| IT Services and IT Consulting | 19% |
| Female | 18% |
| Director / Manager | 18% |
| Firms with 5,000 to 10,000 employees | 18% |

| **Recognition** | |
| --- | --- |
| **Respondent Type** | **Response %** |
| Hospitals and Healthcare | 82% |
| Firms with 5,000 to 10,000 employees | 77% |
| CxO / VP IT / Head of IT | 75% |
| Non-U.S. | 74% |
| Male | 74% |
| Firms with 1,000 to 4,999 employees | 73% |
| U.S. | 72% |
| Firms with more than 10,000 employees | 72% |
| Firms with less than 1,000 employees | 72% |
| Female | 71% |
| Financial Services | 71% |
| Higher Education | 70% |
| IT Services and IT Consulting | 69% |
| CISO | 65% |
| Director / Manager | 64% |
| Government Administration | 64% |
| Best Places | 59% |

# Employee Career Advancement

| Respondent Type | Response % |
| --- | --- |
| Government Administration | 58% |
| Best Places | 53% |
| Hospitals and Healthcare | 53% |
| U.S. | 52% |
| Firms with more than 10,000 employees | 51% |
| Female | 50% |
| Financial Services | 50% |
| Firms with 1,000 to 4,999 employees | 50% |
| Director / Manager | 50% |
| Male | 49% |
| CxO / VP IT / Head of IT | 49% |
| Higher Education | 48% |
| Firms with less than 1,000 employees | 47% |
| CISO | 46% |
| Non-U.S. | 46% |
| Firms with 5,000 to 10,000 employees | 44% |
| IT Services and IT Consulting | 41% |

# Impacting the Profession:

## Deeper & More Diverse Talent Pool Survey

**Survey Question:**

*"Who should technology leaders focus their efforts on to help build a deeper and more diverse talent pool to support the future of the computing profession?" (Multiple selections allowed)*

- Primary and secondary school students (primarily ages 6 to 18).
- College and university students.
- Experienced professionals (in computing and non-computing roles).
- Other (please specify).
- None of the above. This should not be a focus for technology leaders.

## Methodology

- **Total Responses:** 664 technology leaders
- **Geographic Coverage:** 48 countries
- **Survey Method:** Online questionnaire distributed through LinkedIn
- **Response Rate:** 15%

# Respondent Demographics

## By Geographic Region:

- U.S.: 50% (335 respondents)
- Non-U.S.: 50% (329 respondents)

## By Organization Size:

- <1,000 employees: 26%
- 1,000-5,000 employees: 25%
- 5,000-10,000 employees: 13%
- 10,000 employees: 36%

## By Leadership Level:

- C-Suite (CIO, CTO, CDO): 84%
- Director / Manager: 5%
- CISO: 11%

## By Gender:

- Male: 72%
- Female: 28%

## By Industry:

- IT Services and IT Consulting (n=46)
- Financial Services (n=55)
- Government Administration (n=28)
- Hospitals and Health Care (n=53)
- Higher Education (n=37)

# Ranked Results by Respondent Type and Focus Area

| Primary and Secondary School Students (Primarily Ages 6 to 18) | |
| --- | --- |
| **Respondent Type** | **Response %** |
| Director / Manager | 84% |
| Female | 84% |
| Firms with less than 1,000 employees | 80% |
| Higher Education | 78% |
| U.S. | 76% |
| Financial Services | 76% |
| Firms with 1,000 to 4,999 employees | 76% |
| IT Services and IT Consulting | 76% |
| CISO | 76% |
| Non-U.S. | 75% |
| CxO / VP IT / Head IT | 75% |
| Hospitals and Healthcare | 74% |
| Firms with more than 10,000 employees | 73% |
| Male | 73% |
| Firms with 5,000 to 10,000 employees | 70% |
| Government Administration | 61% |

| Respondent Type | Response % |
| --- | --- |
| Higher Education | 89% |
| Hospitals and Healthcare | 81% |
| Government Administration | 79% |
| Director / Manager | 78% |
| Male | 76% |
| Firms with more than 10,000 employees | 76% |
| Non-U.S. | 76% |
| CxO / VP IT / Head IT | 76% |
| Firms with 5,000 to 10,000 employees | 75% |
| U.S. | 75% |
| Female | 74% |
| CISO | 74% |
| Firms with less than 1,000 employees | 74% |
| Firms with 1,000 to 4,999 employees | 73% |
| IT Services and IT Consulting | 70% |
| Financial Services | 67% |

| **Experienced Professionals (In Computing and Non-Computing Roles)** | |
| --- | --- |
| **Respondent Type** | **Response %** |
| Higher Education | 81% |
| IT Services and IT Consulting | 65% |
| CISO | 63% |
| Female | 62% |
| Financial Services | 62% |
| Government Administration | 61% |
| Firms with more than 10,000 employees | 61% |
| Non-U.S. | 59% |
| CxO / VP IT / Head IT | 59% |
| U.S. | 58% |
| Male | 57% |
| Firms with less than 1,000 employees | 57% |
| Hospitals and Healthcare | 57% |
| Firms with 1,000 to 4,999 employees | 56% |
| Firms with 5,000 to 10,000 employees | 53% |
| Director / Manager | 44% |

## Other

| Respondent Type | Response % |
| --- | --- |
| Government Administration | 29% |
| CISO | 27% |
| Financial Services | 25% |
| U.S. | 23% |
| Firms with more than 10,000 employees | 23% |
| Firms with 5,000 to 10,000 employees | 22% |
| IT Services and IT Consulting | 22% |
| Male | 22% |
| Firms with 1,000 to 4,999 employees | 21% |
| CxO / VP IT / Head IT | 20% |
| Director / Manager | 19% |
| Firms with less than 1,000 employees | 18% |
| Non-U.S. | 17% |
| Female | 16% |
| Hospitals and Healthcare | 15% |
| Higher Education | 11% |

| This Should Not be a Focus for Technology Leaders | |
| --- | --- |
| **Respondent Type** | **Response %** |
| Higher Education | 5% |
| Government Administration | 4% |
| CISO | 3% |
| Non-U.S. | 2% |
| Financial Services | 2% |
| Firms with less than 1,000 employees | 1% |
| Firms with 5,000 to 10,000 employees | 1% |
| Male | 1% |
| Firms with 1,000 to 4,999 employees | 1% |
| CxO / VP IT / Head IT | 1% |
| Female | 1% |
| Firms with more than 10,000 employees | 0% |
| U.S. | 0% |
| Hospitals and Healthcare | 0% |
| IT Services and IT Consulting | 0% |
| Director / Manager | 0% |

# Impacting the Profession:
## STEM Career Development Organizations

### Organizations Mentioned in this Book

### Technovation

**Website:** www.technovation.org

**Location:** Global organization based in the United States

**Mission:** Technovation's mission is to empower girls and young women worldwide to learn and use skills needed to solve real-world problems through technology and entrepreneurship. By promoting a culture of innovation, Technovation aims to build a more inclusive future where women play a major role in tackling global challenges through STEM.

### STEM Advantage

**Website:** www.stemadvantage.org

**Location:** Huntington Beach, CA, United States

**Overview:** STEM Advantage was founded in 2012 to mentor, prepare, and inspire students to pursue careers in science, technology, engineering, and math (STEM) through paid internships, mentorships, and scholarships. This 360-degree program is designed to open doors and support students who have talent but don't have the same opportunities and network as students attending a private or elite public university. STEM Advantage partners with students attending the California State University (CSU) system, the largest and most ethnically diverse university system in the nation.

## Women 4 STEM

**Website:** www.women4stem.com.au

**Location:** Australia

**Overview:** Founded in 2005, Women 4 STEM is a not-for-profit organization that operates to increase the entry, retention, and progress of women in the STEM industries. The vision is to actively promote and support women throughout their STEM careers, via a range of practical and sustainable initiatives that will promote urban and regional women's entry, retention, and progression in tech.

## i.c.stars

**Website:** www.icstars.org

**Location:** United States

**Overview:** i.c.stars was established to create economic opportunity for underserved communities by bridging disconnected young adults with the high-growth tech sector. Participants learn by doing; they build web-based applications to solve client challenges, with coding, business, and leadership instruction provided along the way. They also gain the professional network needed to jumpstart their careers.

## Fastrack into Information Technology (FIT)

**Website:** www.fit.ie

**Location:** Ireland

**Overview:** FIT is committed to the socio-economic transformation of marginalized job seekers and disadvantaged communities through innovative ICT training courses as a gateway to employment. It supports jobseekers at risk of or already experiencing long-term unemployment and early school leavers from disadvantaged communities.

# List of Figures


**Figure 1-1:** *The Impact Model of Technology Leadership*...........15

**Figure 3-1:** *Discovery Skills Survey: All Respondents* .............32

**Figure 3-2:** *Discovery Skills Survey: Senior Executives* ........... 32

**Figure 3-3:** *Discovery Skills Survey: CIO 100* ..................... 35

**Figure 4-1:** *Motivation Practices of Technology Leaders*.......... 97

**Figure 4-2:** *Motivation Practices of Technology Leaders: Senior Executives* .............................. 98

**Figure 4-3:** Creating a Great Workplace.......................102

**Figure 4-4:** *Creating a Great Workplace: Factors & Motivators* ...103

**Figure 5-1:** Computer & Information Sciences Bachelor's Degrees Awarded 1971-2022 ....................... 153

**Figure 5-2:** Computer & Information Sciences Bachelor's Degrees Awarded to Females 1971-2022 ........... 157

**Figure 5-3:** Representation by Group in U.S. Population, Bachelor's and Computer Science/Information Science (Cs/IS) Bachelor's Degrees (2022).....................................158

# Works Cited

**Chapter 1**

1. Dyer, Jeffrey H., et al. *The Innovator's DNA: Mastering the Five Skills of Disruptive Innovators.* Harvard Business Review Press, 2019.

**Chapter 2**

1. NASA. "Three Rover Exhibit Opens at Smithsonian's National Air and Space Museum." NASA, 27 March 2015. www.nasa.gov/image-article/three-rover-exhibit-opens-smithsonians-national-air-space-museum-nasm/. Accessed 17 November 2025.

2. Flake, Gary. "Is Pivot a Turning Point for Web Exploration?" TED, February 2010. www.ted.com/talks/gary_flake_is_pivot_a_turning_point_for_web_exploration.

3. Suqi, Rima. "On Your iPhone: Point and Paint." *The New York Times,* 20 May 2009. https://www.nytimes.com/2009/05/21/garden/21apps.html?searchResultPosition=3.

4. Bounds, Gwendolyn. "Narrowing 3,500 Paint Choices to 1 Perfect Hue." *The Wall Street Journal,* 14 April 2011. https://www.wsj.com/articles/SB1000142405274870373010457626100410787343O?gaa_at=eafs&gaa_n=AWEtsqfBB8Rtzq-xTW3oX5X-rVz94ZtbXbp1hSHFQeq5Aizl73Q8YRDlZssV2YgwfvN4%3D&gaa_ts=691b32e1&gaa_sig=KWxJEABW5tvl4uWx52WdAAPnbUT6gso9NsOIk1AA1vVb7doD6SoLrwClZVjdVJAGXjvk4mAW-cfmA7jAUgPfFAw%3D%3D

5. "Olympic Color Genius." *The Webby Awards,* 2013. https://winners.webbyawards.com/2013/websites-and-mobile-sites/general-desktop-mobile-sites/lifestyle/146610/olympic-color-genius. 17 November 2025.

6. *2022 Annual Report.* Otis Worldwide Corporation, 2022. https://s203.q4cdn.com/227649559/files/doc_financials/2022/ar/2022-Otis-Annual-Report.pdf.

## Chapter 6

1. Hopper, Grace Murray. "The Wit and Wisdom of Grace Hopper." OCLC Newsletter, no. 167, Mar./Apr. 1987, www.cs.yale.edu/homes/tap/Files/hopper-wit.html.

2. Herzberg, Frederick. "One More Time: How Do You Motivate Employees?" Harvard Business Review, Vol. 81, No. 1, Jan. 2003.

## Chapter 11

1. Page, Scott E. *The Difference: How the Power of Diversity Creates Better Groups, Firms, Schools, and Societies.* Princeton University Press, 2007.

2. United States Census Bureau. "Number of IT Workers Has Increased Tenfold Since 1970, Census Bureau Reports." Press Release CB16-139, 16 Aug. 2016, www.census.gov/newsroom/archives/2016-pr/cb16-139.html

3. National Center for Education Statistics. "Table 325.35. Degrees in Computer and Information Sciences Conferred by Postsecondary Institutions, by Level of Degree and Sex of Student: Academic Years 1964–65 through 2021–22." U.S. Department of Education, Nov. 2023, https://nces.ed.gov/programs/digest/d23/tables/dt23_325.35.asp

4. U.S. Census Bureau, Population Division. Annual Estimates of the Resident Population by Sex, Race, and Hispanic Origin for the United States: April 1, 2020 to July 1, 2024 (NC-EST2024-SR11H). Release Date: June 2025. https://www2.census.gov/programs-surveys/popest/tables/2020-2024/national/asrh/nc-est2024-sr11h.xlsx

5. "Women Working in Mining in South Africa Speaking Out about the Difficulties." *International Women in Mining,* 1 Dec. 2015, https://internationalwim.org/women-working-in-mining-in-south-africa-speaking-out-about-the-difficulties

6. "CEO Sundar Pichai Speech, Appearance at Girls' Coding Event (Technovation)." The Verge, 10 Aug. 2017, www.theverge.com/2017/8/10/16129490/ceo-sundar-pichai-speech-appearance-girls-coding-event-technovation

7. National Department of Health (NDoH), Statistics South Africa (Stats SA), South African Medical Research Council (SAMRC), and ICF. South Africa Demographic and Health Survey 2016. Pretoria, South Africa, and Rockville, Maryland, USA: NDoH, Stats SA, SAMRC, and ICF, 2019.

8. "Our Alumnae." *Technovation,* https://www.technovation.org/our-alumnae/

9. STEM Advantage, https://stemadvantage.org/

10. President's Council of Advisors on Science and Technology. *Engage to Excel: Producing One Million Additional College Graduates with Degrees in Science, Technology, Engineering, and Mathematics.* Executive Office of the President, Feb. 2012, obamawhitehouse.archives.gov/sites/default/files/microsites/ostp/pcast-engage-to-excel-final_2-25-12.pdf.

11. Women 4 STEM, www.women4stem.com.au/. Accessed 17 Nov. 2025.

12. *Revenge of the Nerds.* Directed by Jeff Kanew, performances by Robert Carradine, Anthony Edwards, Ted McGinley, and Bernie Casey, 20th Century Fox, 1984.

13. *The IT Crowd.* Created by Graham Linehan, performances by Chris O'Dowd, Richard Ayoade, and Katherine Parkinson, Talkback Thames, Channel 4, 2006-2013.

14. i.c.stars, www.icstars.org/. Accessed 17 Nov. 2025.

15. "Fastrack into Information Technology." FIT, fit.ie/. Accessed 17 Nov. 2025.

**Being a Technology Leader, It Really Can Be a Wonderful Life**

1. *It's a Wonderful Life.* Directed by Frank Capra, performances by James Stewart, Donna Reed, Lionel Barrymore, and Henry Travers, Liberty Films, 1946.

**Acknowledgments**

1. *The Holy Bible.* Revised Standard Version, Second Catholic Edition, Ignatius Press, 2002.

# Acknowledgments

Many are the plans in the mind of a man, but it is the
purpose of the Lord that will be established.
**—Proverbs 19:21**

When I retired in 2021, I finally had time to pursue the book I had long imagined. I began by compiling a decade's worth of CIO 100 award winners and senior leaders from organizations recognized as the best places to work in technology. Most were CIOs, each with a story of leadership and innovation, and together they formed the foundation for the conversations I hoped would shape this book.

But having never written a book before, doubt crept in. I often wondered if I had taken on more than I was meant to, and I came close to shelving the project altogether. Then, thanks to divine providence, I heard from Jimmy Berelos. Jimmy and I had met briefly years earlier at a local technology event and hadn't spoken since, yet his message on LinkedIn arrived as if by design. He invited me to discuss an initiative he was starting, and we agreed to meet later that week. Looking back, that unexpected message was the push I needed, the moment that changed everything.

Jimmy was assembling an advisory board for a client developing a diagnostic tool to help IT organizations improve engagement and retention. The project aligned perfectly with the themes I was exploring for the book, so I agreed to join. During our call, he asked what I planned to do in retirement, and I told him about the book and the list of CIOs I had compiled. He wondered who was on the list, and when I mentioned Dick Daniels, Jimmy immediately replied that Dick would also be on the advisory board and offered to introduce us. Thanks to Jimmy, I began conducting interviews in earnest, with Dick as my first interviewee.

This book would not have been possible without the generosity of Dick and the other technology leaders who shared their time, stories, and wisdom, even though they had never met me before. I remain deeply grateful to each of them:

- Dick Daniels
- Vishal Gupta
- Hans Keller
- Barry Lowry
- Cindy McKenzie
- Dr. Philile Mkhize
- Julie Ragland
- Simon Reiter
- Jim Rinaldi
- Graeme Thompson
- Renee Zaugg
- Steve Zerby

I would also like to thank three exceptional technology leaders, John Marcante, Tanya Hannah, and Kirk Kelly, and FIT CEO, Peter Davitt, whose time, insights, and perspectives have helped reinforce and enrich the message of this book.

The PPG stories would not have been possible without the many remarkable colleagues I was privileged to work alongside. In particular, Jeff Lipniskis and Matt Ficco were instrumental in transforming our vision of a digital paint purchase process from concept to reality. Under Scott Sinetar's business leadership, what started as a single slide became a testament to how persistence and collaboration can turn ambitious ideas into lasting innovation.

I am also grateful to my PPG colleagues Joe Menner, Phil Behrens, and Kevan Farley. Their collaboration on several patents, which I was honored to co-invent, exemplifies their dedication to our shared vision of transforming how consumers purchase paint.

I must also acknowledge the more than 1,600 technology leaders worldwide participating in the three surveys that informed this book. None of them knew me personally, yet they took the time to contribute. Their insights added depth and credibility to this work, and I am profoundly thankful for their support.

I'm deeply grateful to my fellow alumni and colleagues at the University of Pittsburgh's School of Computing and Information, especially Mark Shozda, Amelius Carillo, Mackenzie Ball, Terri Taylor, and Melissa Kelly. Collaborating with people so genuinely invested in student success to develop a mentorship program for them reminded me why supporting aspiring professionals in our field matters so much.

Thanks to Julia Poepping and Joe Mertz's support, I had the privilege of working with Carnegie Mellon University students on projects benefiting local non-profit organizations. While I served as their advisor, I learned far more than I taught, particularly about the untapped innovative potential of the next generation of technology professionals.

I must also thank Chris Kowalsky, Steve Agnoli, and David Ulicne from the Carnegie Mellon University Chief Information and Digital Officer (CIDO) program. The opportunity to share my experience as a CIO and the research for this book with aspiring CIDOs has helped reinforce my mission to complete this book.

Finally, I must thank my bride, Janice. You have been by my side for all these years, helping me grow into a better man, husband, father, and grandfather. You supported me throughout my career and stood by me throughout this book-writing journey, encouraging me even when I doubted myself. I am forever grateful for your patience, unwavering belief in me, and enduring love.

# Notes

# Notes

# About the Author

C hris Caruso spent four decades leading technology organizations through periods of rapid change and innovation. As Chief Information Officer at PPG, he led an IT team that earned recognition with the CIO 100 Award for innovation and a spot on Computerworld's "Best Places to Work in IT" list. His work on digital paint color selection tools led to him being a co-inventor on six patents for using technology to assist consumers in the paint color selection and purchasing process.

Now, Chris focuses on what he's learned along the way. He mentors university students and emerging technology leaders, advises nonprofits on their digital strategies, and writes and speaks about the evolving demands of technology leadership. His perspective comes from more than 40 years of experience navigating the challenges technology leaders face daily.